Systems Engineering

I dedicate this book to my entire family.

SCIENCES

Systems and Industrial Engineering,
Field Director – Jean-Paul Bourrières

Industrial and Productive Engineering,
Subject Head – Samir Lamouri

Systems Engineering

Analysis, Modeling and Simulation of Systems

Emmanuel Hygounenc

iSTE

WILEY

First published 2023 in Great Britain and the United States by ISTE Ltd and John Wiley & Sons, Inc.

ISTE Ltd
27-37 St George's Road
London SW19 4EU
UK

www.iste.co.uk

John Wiley & Sons, Inc.
111 River Street
Hoboken, NJ 07030
USA

www.wiley.com

Library of Congress Control Number: 2022948723

British Library Cataloguing-in-Publication Data
A CIP record for this book is available from the British Library
ISBN 978-1-78945-108-5

ERC code:
PE7 Systems and Communication Engineering
PE7_3 Simulation engineering and modelling

Contents

Foreword xv

Acknowledgments xvii

Introduction xix

Part 1. Systems Theory 1

Chapter 1. Systems Theory 3

1.1. The definition of a system 3
1.2. Definition of a complex system 5
1.3. Definition of a system of systems 6
1.4. The systems approach 7
1.4.1. The reductionist approach 7
1.4.2. The holistic approach 8
1.4.3. The analytical and systemic approach 9
1.4.4. The emergence 10
1.5. The systemic method 14
1.5.1. Systemic exploration 15
1.5.2. Qualitative modeling 17
1.5.3. Quantitative modeling 20
1.5.4. The simulation 21
1.6. How to understand the complexity of a system 22
1.6.1. Theme 1: the system is in an environment 23
1.6.2. Theme 2: the system is structured 24
1.6.3. Theme 3: the system evolves over time and has a dynamic 24
1.6.4. Theme 4: the system is controlled 25
1.7. Conclusion on systems theory 31

Part 2. Systems and Requirements Engineering 33

Chapter 2. Introduction to Systems Engineering 35

2.1. The system meets needs. 35
2.1.1. Definition of a need. 35
2.2. Definition of a stakeholder requirement 36
2.3. How to go from need to system 36
2.4. Definition of systems engineering 36
2.5. Iterative systems engineering process 38
2.5.1. Prescription and system construction 38
2.5.2. Needs analysis. 40
2.5.3. Requirements specification 41
2.5.4. Functional and logical design 41
2.5.5. Physical design 42
2.5.6. Verification and validation. 43
2.5.7. Evaluation and comparison 43
2.5.8. The requirements cycle. 44
2.6. System architecture 45
2.7. V-cycle 47
2.8. W-cycle. 48
2.9. Conclusion on systems engineering 49

Chapter 3. Introduction to Requirements Engineering 51

3.1. Definition of requirements engineering 51
3.2. The importance and challenges of requirements engineering 53
3.3. Problem domain and solution domain 54
3.4. Formalizing stakeholder needs and system requirements 55
3.5. Validate system requirements. 56
3.6. Allocate system requirements to subsystems 56
3.7. Validate the allocated requirements of the subsystems. 56
3.8. Choose the solution: define the requirements of the subsystems 56
3.9. Managing requirements 57
3.9.1. Why manage requirements? 57
3.9.2. What are the risks associated with poor requirements management?. . . . 57
3.9.3. Managing requirement attributes 58
3.9.4. Managing the configuration of requirements. 59
3.9.5. Managing changes 61
3.10. Conclusion on requirements engineering. 62

Part 3. Definition of Requirements 65

Chapter 4. Unifying Thread Example 67

4.1. Objective of the unifying thread example 67
4.2. Presentation of the raw specifications 68

Chapter 5. Needs Analysis 71

5.1. Objectives of needs analysis 71
5.2. Stakeholder identification 72
5.2.1. Definition of stakeholder 72
5.2.2. PESTEL analysis 73
5.2.3. Analysis by type of environment 74
5.2.4. Application to the unifying thread example: stakeholder identification 76
5.3. Identification of external interactions 78
5.3.1. Objectives of identifying external interactions 78
5.3.2. Application to the unifying thread example: identification of external interactions 80
5.4. Collection of needs 83
5.4.1. Why capture needs? 83
5.4.2. Methods for collecting needs 84
5.4.3. Application to the unifying thread example: formalization of needs 85
5.5. Identification of the life cycle 87
5.5.1. Building the system life cycle 87
5.5.2. Application to the unifying thread example: identifying the life cycle . . . 89
5.6. Identification of the system's missions 91
5.6.1. Definition of the system's missions 91
5.6.2. Application to the unifying thread example: mission definition 94
5.7. Identification of operational scenarios 96
5.7.1. Defining an operational scenario 96
5.7.2. Application to the unifying thread example: definition of operational scenarios 97
5.8. Identification of services and constraints 100
5.8.1. From missions to services 100
5.8.2. Application to the unifying thread example: identification of services and constraints 101
5.9. Formalization of stakeholder needs 102
5.9.1. The input specifications 102
5.9.2. Application to the unifying thread example: formalization of needs 107
5.10. Conclusion on needs analysis 110

Chapter 6. Requirements Specification 111

6.1. Objective of the requirements specification process 111
6.2. Identification of the system's functional modes. 113
6.2.1. Definition and purpose of the functional modes 113
6.2.2. What to do to identify functional modes 114
6.2.3. Application to the unifying thread example: identifying the functional modes 115
6.4. Identification of system functions 116
6.4.1. What is a system function?. 116
6.4.2. What to do to identify the system functions 119
6.4.3. Application to the unifying thread example: identification of system functions 122
6.5. Identification of external interactions 123
6.6. Defining system behaviors 124
6.6.1. Objective of defining the functional behavior of the system 124
6.6.2. Application to the unifying thread example: definition of a functional scenario 124
6.7. Defining the system requirements 126
6.7.1. Why define the system requirements?. 126
6.7.2. What to do to define the system requirements 128
6.7.3. How to define system requirements: application to the unifying thread example. 130
6.8. System specification. 134
6.8.1. The specification document 134
6.8.2. Application to the unifying thread example: system specification 138
6.9. Conclusion on requirements specification. 140

Chapter 7. Requirements Validation 143

7.1. General process 144
7.2. Selecting methods and defining validation procedures. 145
7.3. Establishing requirements traceability 145
7.3.1. Establishing traceability from needs to requirements 145
7.3.2. Establish traceability from requirements to needs 145
7.4. Analysis of assumptions and induced requirements 146
7.4.1. Analysis of assumptions 146
7.4.2. Analysis of induced requirements 146
7.5. Rolling out the validation 146
7.5.1. Checking the requirements individually (correctness). 147
7.5.2. Globally checking the requirements (consistency). 147
7.5.3. Validating requirements against needs 148

7.6. Identifying and resolving gaps between needs and requirements 148
7.6.1. Identifying gaps and conflicts 148
7.6.2. Assessing conflicts 148
7.6.3. Study of trade-offs 149
7.6.4. Identifying trade-offs and impacts. 149
7.7. Saving a database of validated requirements 149
7.8. Conclusion on requirements validation 150

Part 4. System Design 151

Chapter 8. Functional and Logical Design 153

8.1. Design and functional architecture 153
8.1.1. Static functional architecture. 153
8.1.2. Dynamic functional architecture. 154
8.1.3. Behavioral functional architecture. 156
8.2. Identifying the sub-functions of the system and their interactions. 157
8.2.1. Purpose of identifying sub-functions 157
8.2.2. Functional chain analysis approach 158
8.2.3. Implementation of the unifying thread example: identification of sub-functions 161
8.2.4. The outputs of the static architecture 164
8.2.5. Rules on the quality of a functional decomposition 167
8.3. Functional interface analysis 168
8.3.1. The coupling matrices 168
8.3.2. Characterizing functional interfaces. 171
8.4. Consolidation of functional modes. 171
8.4.1. Objective of consolidating functional modes. 171
8.4.2. What to do to identify the functional sub-modes. 172
8.4.3. Application to the unifying thread example: resumption of the functional modes. 174
8.5. Resuming system functioning 176
8.5.1. Objective of the system recovery 176
8.5.2. Implementation of the unifying thread example: resumption of the system functioning. 178
8.6. Propose groupings of sub-functions into logical components 180
8.6.1. The use of logical components. 180
8.6.2. Proposing a schematic diagram of the solution. 180
8.7. Allocate functional requirements to sub-functions 180
8.7.1. Objective of the allocation of system requirements to sub-functions. . . . 180
8.7.2. Application to the unifying thread example: allocation of system requirements to sub-functions 182
8.8. Conclusion on functional design 184

Chapter 9. Verification and Validation of the Functional Architecture . . . 185

9.1. Verification of the functional architecture 185
9.2. Validation of functional architecture 186
9.2.1. Selecting methods and defining validation procedures 187
9.2.2. Establishing traceability 187
9.2.3. Analysis of assumptions 187
9.2.4. Unrolling the validation 188
9.2.5. Performing the revalidation 188
9.2.6. Recording the results of the functional architecture's validation 188
9.3. Conclusion on the verification and validation of functional design 188

Chapter 10. Physical Design 191

10.1. Purpose of physical design 191
10.2. Identification of physical components 193
10.2.1. Identifying components and allocating technical functions 194
10.2.2. Completing the identification of the components with the missions of the system 195
10.2.3. Completing the identification of components with non-functional requirements 195
10.2.4. Identification of variants in the physical architecture 195
10.2.5. Building physical architectures with a product line approach 196
10.2.6. Building physical architectures from a logical architecture 197
10.2.7. Application to the unifying thread example: identification of components 198
10.3. Decomposition of the components 201
10.3.1. Architecture principles 202
10.3.2. Coupling matrix 205
10.3.3. Global coupling quality of an architecture 206
10.3.4. The physical tree structure 209
10.3.5. The architecture of the components 211
10.3.6. Application to the unifying thread example: components and interfaces . 212
10.4. Characterization of interfaces 215
10.5. Identification of system configurations 218
10.5.1. Definition of a technical configuration 218
10.5.2. Activities to identify technical configurations 219
10.5.3. Application to the unifying thread example: identification of configurations 220
10.6. Identifying the physical functioning of the system 222
10.6.1. Definition of the physical functioning of the system 222
10.6.2. Application to the unifying thread example: constructional scenario . . . 223
10.7. Allocation and definition of subsystem requirements 224
10.7.1. Emergence of properties 224

10.7.2. Architecture sizing 226
10.7.3. Allocating non-functional requirements 227
10.7.4. Definition of subsystem requirements 228
10.8. Conclusion on the physical design 230

Chapter 11. Verification and Validation of the Physical Architecture 231

11.1. Verification of the physical architecture 231
11.1.1. Defining physical design verification procedures. 232
11.1.2. Performing physical design verification. 232
11.1.3. Rechecking the physical design 233
11.1.4. Recording the results of the physical design verification 233
11.2. Validation of the physical architecture 233
11.3. Conclusion on the verification and validation of the physical architecture . . 235

Chapter 12. Evaluation and Comparison of Solutions. 237

12.1. Evaluation of architectures 237
12.1.1. Selection of common criteria. 237
12.1.2. Value criteria. 237
12.1.3. Risk and cost criteria 238
12.2. Comparison of architectures. 239
12.2.1. Selecting stakeholders. 239
12.2.2. Characterizing the weight of each criterion 239
12.2.3. Getting stakeholders to vote 240
12.2.4. Comparing architectures 240
12.2.5. Justifying architectural choices. 241
12.3. Conclusion on the evaluation and comparison of the architectures 241

Part 5. Virtual System Integration 243

Chapter 13. Integration on a W-Cycle 245

13.1. Virtual integration and real integration. 245
13.2. Simulation models 247
13.3. Conclusion on integration on a W-cycle 249

Chapter 14. Creating a Simulable Design Model 251

14.1. Defining the simulation objectives 251
14.2. Simulation and the systemic approach 251
14.3. Analysis and modeling of system architecture. 252
14.4. Analysis and causal modeling 254
14.4.1. Extracting simulation variables. 254
14.4.2. Relationships between variables 255

14.4.3. Feedback loops . . . 255
14.4.4. Application to the unifying thread example: causal analysis . . . 258
14.5. Formalizing the simulation model . . . 261
14.5.1. Classification of variables . . . 261
14.5.2. Dynamic modeling . . . 262
14.5.3. Application to the unifying thread example: formalization of the causal model . . . 265
14.6. Running the simulation . . . 266
14.6.1. Identifying and configuring equations . . . 266
14.6.2. Configuring the simulation . . . 266
14.6.3. Application to the unifying thread example: simulation . . . 266
14.7. Analyzing the results . . . 268
14.7.1. Purpose of the analysis of the results . . . 268
14.7.2. Application to the unifying thread example: analysis of the results . . . 268
14.8. Conclusion on the realization of a simulable design model . . . 269

Chapter 15. Making a Simulable Specification Model . . . 271

15.1. The monitoring model . . . 271
15.1.1. Why make a monitoring model? . . . 271
15.1.2. How to build a monitoring model . . . 272
15.2. Writing structured requirements . . . 273
15.2.1. Reminder on the structure of a requirement . . . 273
15.2.2. Application to the unifying thread example: selecting structured requirements . . . 274
15.3. Formalizing property requirements . . . 274
15.3.1. Properties . . . 274
15.3.2. How to formalize a property . . . 275
15.3.3. Application to the unifying thread example: formalization of properties . . . 276
15.4. Building a monitoring model . . . 278
15.4.1. Application to a property . . . 278
15.4.2. Creating the specification model . . . 279
15.5. Running the simulation . . . 280
15.5.1. Connecting the specification model . . . 280
15.5.2. Application to the unifying thread example: running the specification model . . . 281
15.6. Analyzing the results . . . 283
15.6.1. Result from the specification model . . . 283
15.6.2. Application to the unifying thread example . . . 283
15.7. Conclusion on the realization of a simulable specification model . . . 288

Part 6. System Integration 289

Chapter 16. Using Models 291

16.1. Models in systems engineering 291
16.2. Typology of models 291
16.3. Analytical behavioral models 292
16.3.1. Using the behavioral model in V&V 293
16.3.2. Model-based design 293
16.4. Business analytical models 296
16.4.1. Using the business model in V&V 296
16.5. Conclusion on the use of models 296

Chapter 17. System Integration, Verification and Validation 297

17.1. System integration 297
17.1.1. Receiving subsystems 298
17.1.2. Assembling subsystems 299
17.2. Checking the system 299
17.2.1. Defining the system verification strategy 300
17.2.2. Defining the system verification procedures 302
17.2.3. Performing the system check 303
17.2.4. Establishing system compliance 304
17.3. System validation 304
17.3.1. Preparing for validation 305
17.3.2. Performing validation 306
17.3.3. Analyzing the results 307
17.3.4. Saving validation results 307
17.4. Conclusion on system integration, verification and validation 308

General Conclusion 309

References 311

Index 315

Foreword

To meet the challenges of our time, every industry must constantly improve, and for this, even though some sophisticated innovations allow for simple answers to seemingly complex challenges, the basic trend is to know how to manage the complexity born from the assembly of many subsystems and components that must be skillfully linked. The concept of a system is therefore increasingly important, whether for systems that are developed to be made available to customers (e.g. aircraft in the aeronautics industry), or for the underlying systems that support an industry (e.g. factories and the entire supply chain, the information systems of a company).

Systems engineering has thus become a fundamental discipline to address the growing complexity of these systems.

It is a broad, multifaceted discipline that has been considerably enriched recently to meet the challenges of the times. This book addresses all these riches.

I personally find in this discipline the strength to allow the passage of common sense on a large scale. This is also its weakness, because this appearance of evidence on a small scale hinders its appropriation and its valorization. In practice, this turns out to be much more difficult than it seems at first sight, and absolutely critical. This is what the retrospective analysis of errors and failures in various complex systems (aeronautics, space, nuclear, etc.) reminds us. Risks come from systemic effects that create complexity from simple elements that interact.

In the aeronautics industry in particular, the one I know best, systems engineering is no longer reserved for aircraft manufacturers who manage the entire aircraft (or helicopter). It has also become fundamental for the largest aeronautical equipment manufacturers, because the systems they develop also require it.

I personally pay particular attention to the concepts of virtual integration. They have become necessary at more and more levels and increasingly regular frequencies to check the solidity of the foundations and second floors during the development. This is an essential issue in connection with the digital transformation and the introduction of agile methods into these developments. Vast fields of progress are still being explored.

Emmanuel plays a key role in appropriating systems engineering at Safran. His multiple competences addressing fundamental, methodological and practical aspects enable him to have a precise and concrete view of this discipline. His experience and this broad vision allow him to propose this precious book to address this rich and fulfilling discipline.

Gildas GARNIER
Director of Engineering 4.0 at Safran
November 2022

Acknowledgments

This book was born at the end of the first lockdown of 2020 with the intention that it be a working document rather than a thesis.

I want to thank Gildas Garnier, Engineering 4.0 Director, for agreeing to write the foreword of my book. I am delighted to be working with him on this major project.

I am very grateful to the group that reviewed my book between two lockdowns, set up by the Safran System Committee and those in charge of the expert network. Thank you very much, this help was precious and allowed me to enrich my content.

I would also like to thank Eric Clavé, my colleague and partner in this fantastic adventure around the world of systems engineering. Without him, it is clear that I would never have gone this far.

My gratitude goes to my colleagues who throughout these years have created a fertile ground that has enabled me to progress and grow in systems engineering. Special thanks to Nicolas, Jean-Marc, Mickael, Théo and Camille.

Finally, I would like to thank the ENI of Tarbes, with whom I have been collaborating for almost 9 years, for allowing me to teach systems engineering, especially Bernard Archimède and Xavier Desforges, as well as Jean-Yves Fourquet for allowing me to give my first lectures.

My thoughts go to my loving wife Cécile, whom I thank for her support, not forgetting my son Clément and our forthcoming new arrival (my daughter Claire).

Introduction

Objectives

This book is an illustrated presentation of the systems and requirements engineering approach that covers the entire development cycle, with a section devoted to systems integration. It is based on a varied bibliography and aims to provide a coherent synthesis of all these references. In addition, this book introduces a recent approach to integration based on simulation models, which we will call virtual integration.

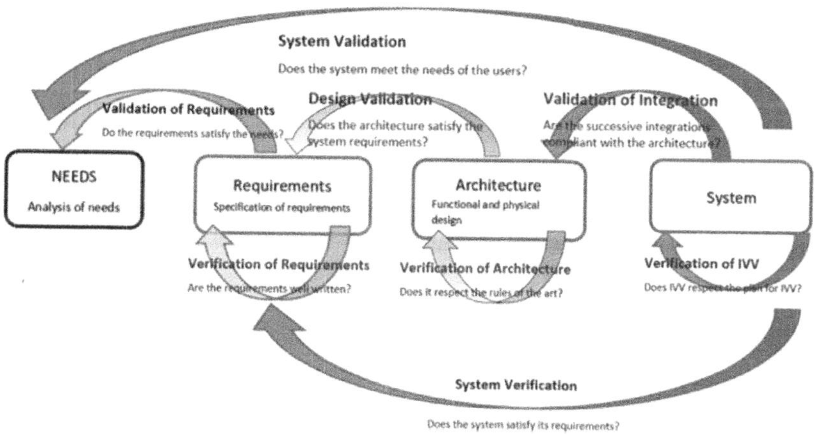

Figure I.1. *Development cycle (AFIS n.d.)*

For a color version of all the figures in this chapter, see www.iste.co.uk/hygounenc/system.zip.

In Figure I.1, we have a generic development cycle of a technological system decomposed into four stages. First, we have the needs analysis, which is then translated into requirements in a specification. Then, we have the architecture phase with the two complementary design steps, the functional and the physical design. We end with the system, which is the final product, the result of integrating a set of components identified during the design. This cycle is not linear, as it consists of many verification and validation loops. Each step has an internal verification loop and a validation loop with the previous step. Verification examines whether the activity is in accordance with the rules of the art, and validation examines whether the activity covers the previous activity. In the aeronautics domain, when implementing the recommendations of ARP 4754, the requirements verification activity does not appear, because it is natively integrated into the requirements validation activity. At the system level, we have a verification loop on the system requirements, i.e. does the system work, and a big validation loop on the needs, i.e. did we make the right system.

The organization of this book enables the reader to approach the design and integration of systems from several points of view, namely:

– systems engineering;

– requirements engineering;

– virtual integration;

– system integration.

Systems engineering point of view

This book enables the reader to approach design from a systems engineering perspective. From this perspective, we will address all phases of a development cycle, namely:

– the definition of the system requirements with:

- stakeholder requirements analysis to identify needs,

- the specification of the technical requirements of the system, which poses the problem to be solved;

– the design of the system with:

- functional design that analyzes the system from an abstract perspective and defines its ability to "do" things,

- physical design that makes it possible to analyze the system under a concrete angle and to define its capacity to "be" something;

– the evaluation and comparison of candidate solutions to select the best architecture based on common value and risk criteria;

– checking and validating the system to determine whether we have made a system that works and meets the needs.

Apart from evaluation and comparison, we find the different steps that we have listed in Figure I.1.

Requirements engineering point of view

This book enables the reader to approach design from a requirements engineering perspective. Although this activity is integrated into the systems engineering process, it can be presented independently.

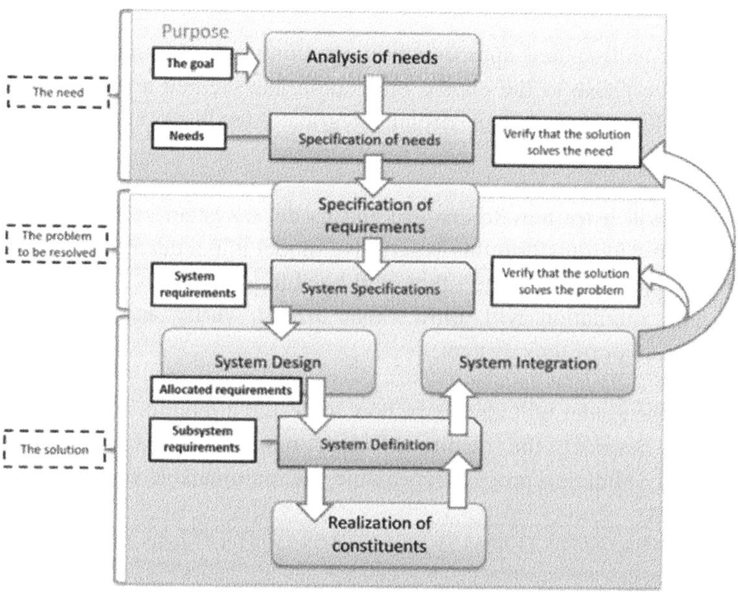

Figure I.2. *Development cycle and requirements engineering*

We will address all the phases of a development cycle from a requirements engineering point of view to include the following.

The development of the requirements with (see Figure I.2):

– the development of requirements from a needs perspective through the needs analysis process;

– requirements development from a solution perspective with:

- the process of specifying requirements, which allows us to pose the problem to be solved,

- functional and physical design processes that allow requirements to be allocated to the lower system level so that subsystem requirements can be defined.

Requirements management with:

– configuration management of requirements;

– change management of requirements.

Virtual integration point of view

This book provides a new approach to integration using only simulatable models. This intermediate phase to the classic integration has secured a project as soon as possible by making intermediate virtual integration feedback in the development cycle.

The reader will learn how to analyze and model a system in a qualitative way. They will see how to transform this descriptive model into a quantitative model and, by introducing the variable "time", they will be able to formalize a set of simulable equations. The simulation will allow them to test, verify and validate several scenarios of behavior of their system.

In parallel, the reader will also learn how to design a monitoring model that will allow them to improve the quality of their requirements and to make their verification and validation process repeatable, i.e. automatable on their simulation and test facilities.

System integration point of view

This book enables the reader to approach the different stages of classical system integration to go from the realization of the components to the finished product. The goal is to verify that the system meets its requirements and to validate that the needs of the environment are covered by carrying out its operational missions.

Organization of the book

The book is structured around six parts. We have two introductory parts that present the origins of systems engineering with systems theory and an introduction to systems engineering through processes. The next four parts follow a W-shaped development cycle, which we will present later and which is an evolution of the classical V-shaped development cycle (see Figure I.3).

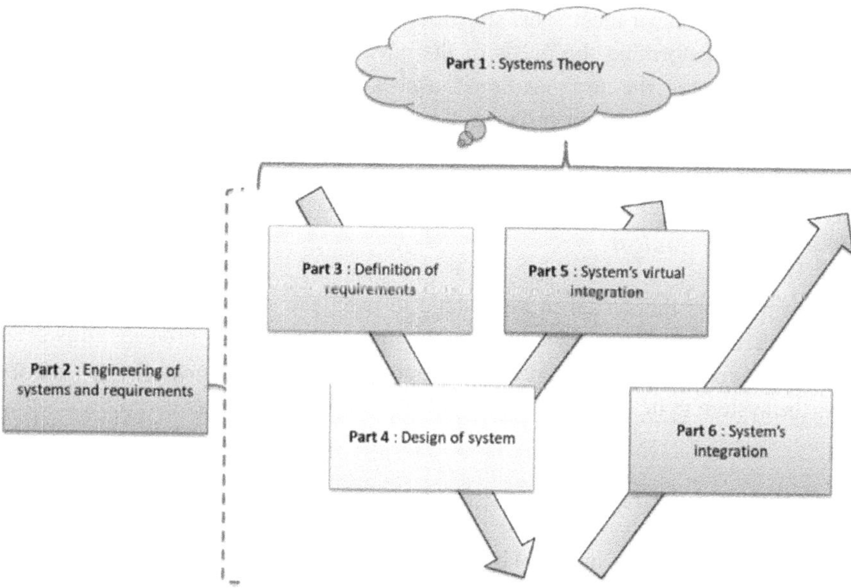

Figure I.3. *Organization of the book*

Part 1: Systems theory

Part 1 is an introduction to systems theory. We will define the notion of system and more particularly complex systems. We will see how to approach them and the properties that characterize them.

Part 2: Introduction to systems and requirements engineering

Part 2 is an introduction to systems and requirements engineering. This section presents the two design perspectives discussed in this book. We will define what systems engineering is and how it can be implemented using the system architecture.

We will present how requirements engineering activities relate to systems engineering. Indeed, even if the requirements engineering activities are part of systems engineering, they can be performed in parallel and presented independently.

Part 3: The definition of the requirements

Part 3 presents the first step in the design of a system, which is the definition of the requirements it must meet. This step allows us to properly define the problem to be solved by transforming the needs of the environment into system requirements. This part integrates the organizational aspect by separating the processes and concepts that concern the project owner, i.e. the client, and the project manager, who is the designer of the solution.

Part 4: System design

Part 4 presents the system design stage to meet the requirements. For this, we will present the two design processes. The functional design process allows the system to be architected from an abstract perspective. The physical design process allows the system to be architected from a concrete angle.

Part 5: Virtual integration

Part 5 is an evolution in the classical realization of the systems. It introduces a backward development cycle approach based on the integration of simulation models. This innovative approach uses two types of models, namely a simulatable design model based on system dynamics and a simulatable specification model that allows for the verification of properties.

Part 6: System integration

Part 6 presents the classical integration of a system. We will discuss partial system integration approaches such as SIL (software-in-the-loop) and HIL (hardware-in-the-loop). We will study the use of analytical and business models for these integration phases. We will present how to proceed with the verification and validation of the system.

PART 1

Systems Theory

1

Systems Theory

This chapter is an introduction to systems or systemic theory. The objective is to present the foundations of this theory, which are based on the notion of system. Indeed, systemic theory is based on the premise that everything can be considered a **system**, i.e. a complex network of interacting elements. The systemic approach is deeply conceptual, and we will see that the first concept is the notion of **totality** where the "whole" is more than the sum of the parts. Specifically, this means that the properties of an integrated system are not limited to the sum of the properties of the elements that constitute it. We also refer to a global or **holistic** approach, which is the basis of systemic theory. With this global approach, we seek to model the emergent properties of the system that arise from the complex cause-and-effect interactions between these elements. These interactions can be looped and provide **feedback** that may be stable or unstable. We also address **circular causality**.

1.1. The definition of a system

We will define a system in several ways, namely in a formal and normative way, in order to lay the theoretical foundations that will allow us to approach the next chapters of this book.

In a **formal way**, in Krob (2014), a system is characterized by (see Figure 1.4):

– a set of input flows X, output flows Y and internal states Q;

– as well as behaviors that link these inputs, outputs and states together over time.

For a color version of all the figures in this chapter, see www.iste.co.uk/hygounenc/system.zip

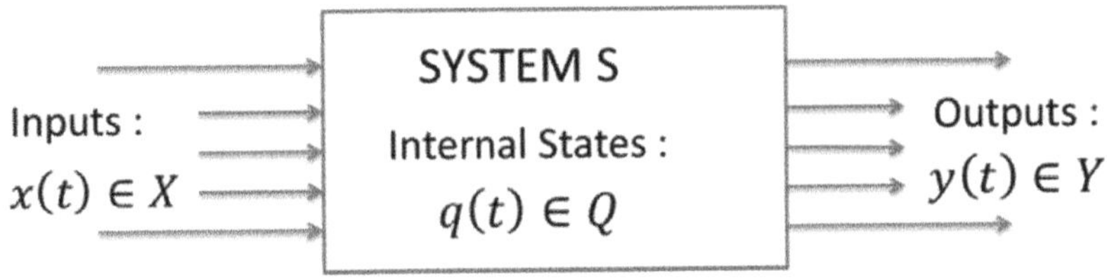

Figure 1.1. *A formal system (Krob 2014)*

Among these behaviors, we have:

– a functional behavior that produces outputs y(t) belonging to Y from inputs x(t) belonging to X, from the current state q(t) belonging to Q;

– an internal behavior that changes the internal state q(t) belonging to Q over time under the action of an input x(t) belonging to X.

In a **standardized way**, a system is a combination of interacting elements organized to achieve one or more defined goals in an environment (ISO/IEC-15288: 2015 – see ISO (2015)).

In a **detailed approach**, a system is (Pollet 2016):

– an organized set of software, hardware and human components in mutual interaction;

– placed in a given environment: the environment consists of other systems;

– fulfilling a certain purpose: the system has a purpose;

– evolving over time to adapt to the loading of the environment and ensure this purpose.

This definition allows us to introduce the analysis of systems from two complementary points of view, which are the **black box** point of view and the **white box** point of view. Each of these views is studied from a **structural**, **temporal** and **steering** aspect (AFIS 2012).

We find in this definition the postulate that everything is a "system" (see Figure 1.5).

However, we will distinguish three families of systems (Meadows 2008; Hoarau 2010):

– **simple systems**: these are linear systems whose behavior is easily predictable;

– **complicated systems**: a set of simple systems that can be understood analytically;

– **complex systems**: these are nonlinear systems whose behavior is difficult to predict. They must therefore be understood as a whole and not by analytical decomposition.

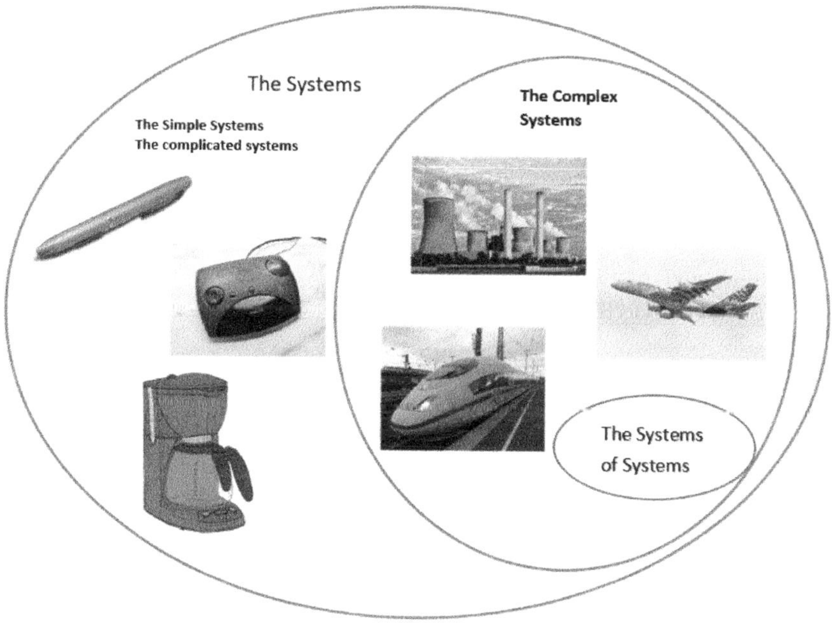

Figure 1.2. *Everything is a "system"*

1.2. Definition of a complex system

To understand what a complex system is, we will compare it to a complicated system based on the work of Le Moigne (1999). A **complicated system** can be explained. Indeed, we can simplify it to discover its intelligibility in the sense of "**explanation**". A **complex system** can also then be understood. Indeed, we must model it to build its intelligibility in the sense of "**understanding**". This is not a definition of the complexity of a system, but it allows us to highlight the importance of modeling a complex system to understand it. However, a link can be made between the complexity of a system and the number of its internal and external interfaces. This increasing complexity is related to several parameters such as:

– the evolution of needs, which has led to an increase in the number of functions and therefore the number of components to be integrated;

– the evolution of technology;

– architecture choices: modular system or integrated system;

– the appearance of systems with leading software;

– critical systems (Cressent 2012).

In Meadows (2008), the author proposes a definition of complex systems as being nonlinear and difficult to predict.

*Cause-and-effect links: indeed, a complex system is made of a set of **networks of interactions** with many **cause-and-effect** links.*

*Deadlines and delays: it has **deadlines.** This means that causes can have short-term effects or long-term effects.*

*Feedback loops: in interactions, it has **feedback loops**. These loops can be **negative**, i.e. they regulate the system by stabilizing it, or they can be **positive**, namely they amplify the effects leading to a divergence.*

Emergent properties: the complex system has emergent properties, i.e. global properties that emerge from the interactions between its constituents. *That is, the "whole", i.e. the integrated system is more than the sum of its parts, its constituents.*

Threshold effects: the complex system may have threshold factors. This means that there are thresholds above which the effects are different from those below the same thresholds.

Key factors: the complex system may have key factors. These are variables for which one cause can lead to great effects. This is what we call the "butterfly" effect.

1.3. Definition of a system of systems

Among the complex systems, we have a particular family that includes the systems of systems. The definition we can use is the following. It is a system that is composed of autonomous systems, i.e. systems that have their own objectives, and that collaborate with each other in order to meet a new need.

Figure 1.3. *The Naval Aviation group: a system of systems*

An example, which illustrates this definition, is the French Navy's Naval Aviation group. Indeed, we have an anti-aircraft frigate, two anti-submarine warfare frigates and an SNA (nuclear attack submarine) that work together to form a new system whose mission is to defend the flagship, the aircraft carrier Charles de Gaulle (see Figure 1.3).

1.4. The systems approach

1.4.1. *The reductionist approach*

Engineering is made up of professions, which are called "disciplines", for example, electrical engineering, mechanical engineering or computer engineering. The discipline approach is a specialist approach that is necessary but it gives a very "disabling" view of the system. This simplifying or reductive approach analyzes only a part of the system or at least according to a certain point of view, which depends on the discipline. This bottom-up view is called **reductionist**.

The reductionist vision is at the origin of the **analytical approach** which studies the system by breaking it down into increasingly detailed elements. This approach is perfectly adapted to systems, which can be decomposed into simple components without loss of information on their functioning (Meinadier 1998). However, in order to be applicable, the analytical approach makes two assumptions for the summability of the components to be possible. The first is that the interactions between the elements analyzed are negligible, and the second is that the relationships that express the behaviors of the elements are linear (von Bertalanffy 1968).

1.4.2. *The holistic approach*

The system engineer in charge of design must rely on this set of skills, but they must also have a more global vision of the product to be made. This approach allows the engineer to study the system in its environment and its external and internal interactions. The importance of such an approach lies in one of the fundamental properties of a system, which is its non-summability. Indeed, we cannot summarize the properties of a system to the sum of the properties of its elements. This global vision, from the top or "top down", is called **holistic**.

The **systemic approach** is based on this holistic vision. It makes it possible to understand complex systems for which the only analytical approach cannot be adapted because it is too simplistic. Indeed, the analytical approach does not allow us to observe all the causal links between the system itself and its environment.

Since a complex system cannot be analyzed simply, the systems approach proposes to understand its behavior in its entirety by modeling and simulating it (Meinadier 1998). This is what Forrester's system dynamics proposes in Forrester (1968), where the complex behaviors of a system are modeled by identifying and interacting with level (state) and flow variables.

Another characteristic of the analytical approach is the study of phenomena by seeking to identify causes and effects. This is called linear causality. This simplifying vision does not capture the interactions that can occur between effects and causes. This is what is called circular causality, which is at the origin of feedback loops and the complexity of interactions (Aracil 1984). Forrester (2009) explains the importance of feedback loops in the nature of systems. In Figure 1.4, we have an initial linear reasoning where we have an observation of a problem which leads to the decision of an action that will lead to a result. This is called an open loop. On the second reasoning, we have the same sequence but with a feedback loop of the results on the information of the problem. This looping will lead to an adjustment of the action which will modify the result and so on. This is what we call the closed loop. For a complex system, it is the study of all its feedback loops that allows us to understand its behavior in a global way.

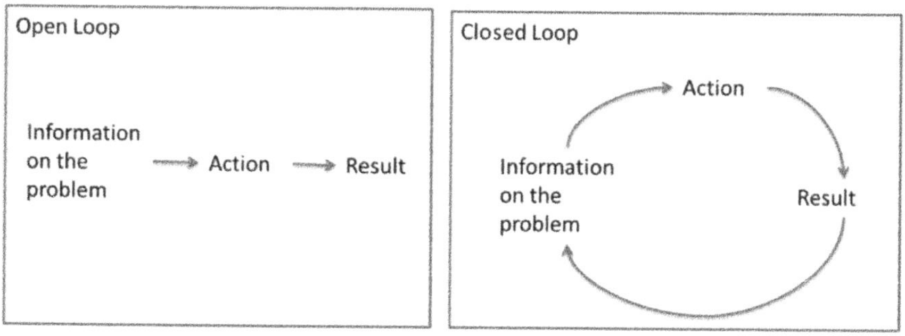

Figure 1.4. *Open loop and closed loop (Forrester 2009)*

1.4.3. *The analytical and systemic approach*

The table in Figure 1.5, taken from Mathelot (1988), presents a comparative synthesis between the analytical approach and the systems approach.

Analytical approach	Systemic approach
1. Reductionist	1. Global (holistic)
2. Isolated	2. Joined
3. Is interested in the nature of interactions	3. Is interested in the effects of interactions
4. With a study by each variable	4. With a study by each group of variables
5. Resulting in the validation from a theory	5. Resulting in the validation from a comparison with reality
6. Researching the evidence	6. Researching the relevance
7. By discipline	7. Multi-disciplinary and trans-disciplinary

Figure 1.5. *Comparison of analytical and systemic approaches*

We will see later on that systems engineering is the integration of the two approaches, or as Thomé (1993) puts it, it is the application of the systemic approach to engineering of the system, which is the set of skills implemented for the design of a complex system (Figure 1.6).

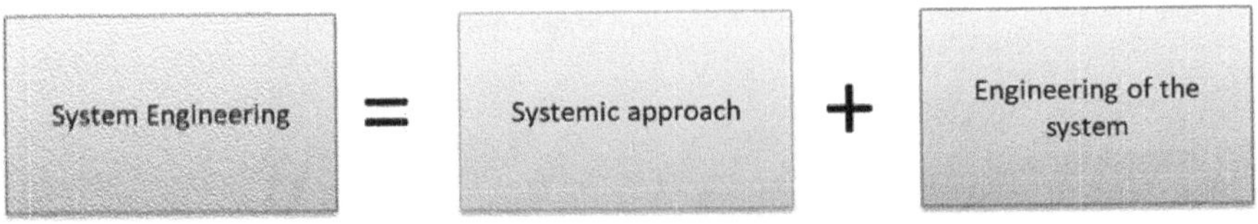

Figure 1.6. *Systems engineering*

Indeed, the systemic approach or the global approach becomes necessary for complex systems because of their heterogeneity. A complex system is heterogeneous because it is made up of matter, informational and human components with numerous cause-and-effect links that loop back and forth. The analytical approach applies well to hardware systems. However, it is more difficult to implement for information systems that interact with humans and human systems in a general way.

1.4.4. *The emergence*

One of the fundamental properties of a system is that "**the whole is more than the sum of its parts**". That is, the properties of an integrated system cannot be deduced from the properties of its subsystems alone. For this, it is necessary to take into account the structure, i.e. all the interactions of the system.

Krob proposes a formal definition of integration that allows the principle of emergence to be introduced (Krob 2014). A formal system S is the result of the integration of its subsystems if there is a formal system C made of the set of these subsystems and a dual mechanism where we have (see Figure 1.7):

– the S system which is the abstraction of the C system;

– the C system which is the concretization of the S system.

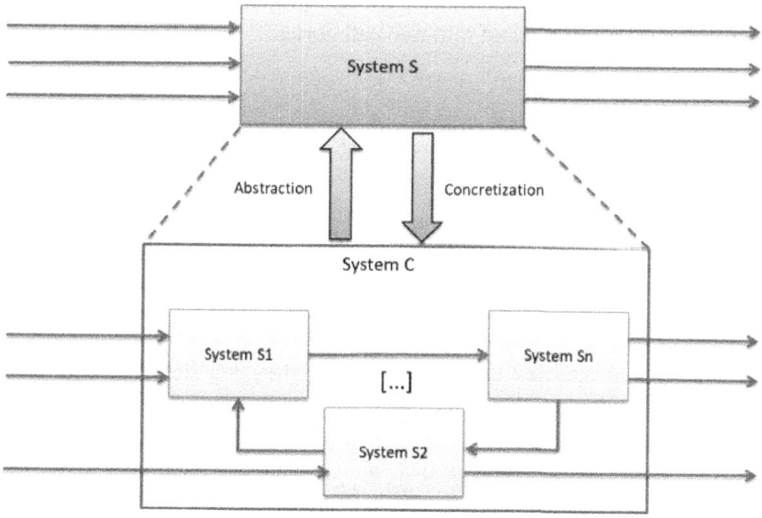

Figure 1.7. *Formal integration of a system (Krob 2014)*

This means that an integrated system is not the simple result of the sum of the models of each subsystem that makes it up. We need two representations to capture the emergence, namely:

– the set of models of the subsystems that make up the system;

– a high-level model representing the system.

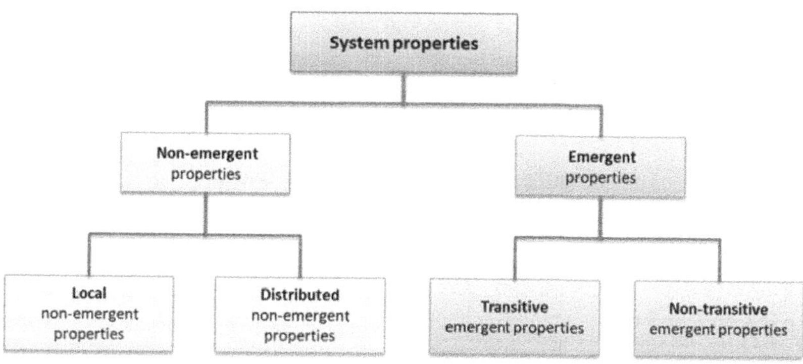

Figure 1.8. *Properties of a system (Thomé 1993)*

In Thomé (1993), a classification of system properties is proposed (see Figure 1.8). These properties are divided into two categories:

Non-emergent properties:

– the non-emergent properties are local, i.e. they are found on a single subsystem;

– distributed non-emergent properties, i.e. those that are distributed over several subsystems.

Emergent properties:

– the emergent properties are transitive, i.e. they are distributed over several subsystems and thus knowledge of the interaction network is necessary to know the properties at the system level;

– non-transitive emergent properties, i.e. properties that depend only on knowledge of the network of interactions of the subsystems to know the properties at the system level.

Emergent and non-emergent properties can be separated into four classes or types (Thomé 1993):

– **constant** properties: properties that do not change over time. For example, "the system must have a length Lt of 10 m". This is a constant and emergent property. Let us say that the system in question is an assembly of three elementary bricks (see Figure 1.9).

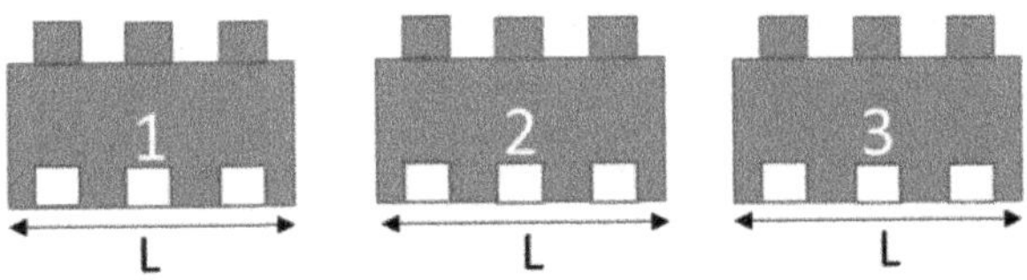

Figure 1.9. *Three elementary bricks*

Each brick has the following properties:

– a length L = 5 m;

– three type 1 and three type 2 interfaces;

– a link between two bricks can be made between a type 1 interface and a type 2 interface (see Figure 1.10).

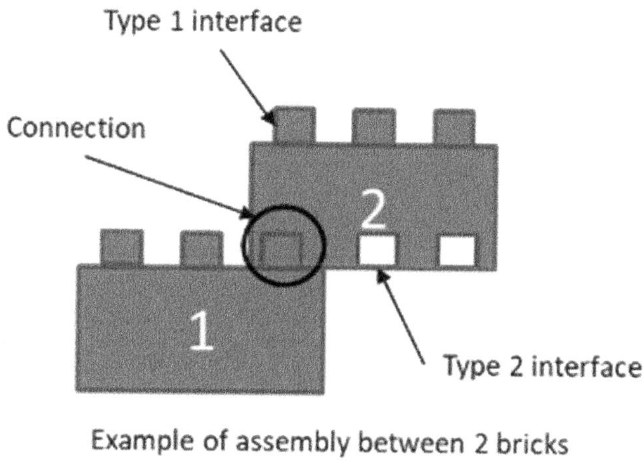

Figure 1.10. *Assembly*

There are three possible assemblies forming three different interaction networks (see Figure 1.11). Only we can meet the requirement for the system to be 10 m long. In Thomé's classification, we are in the presence of a transitive emergent property.

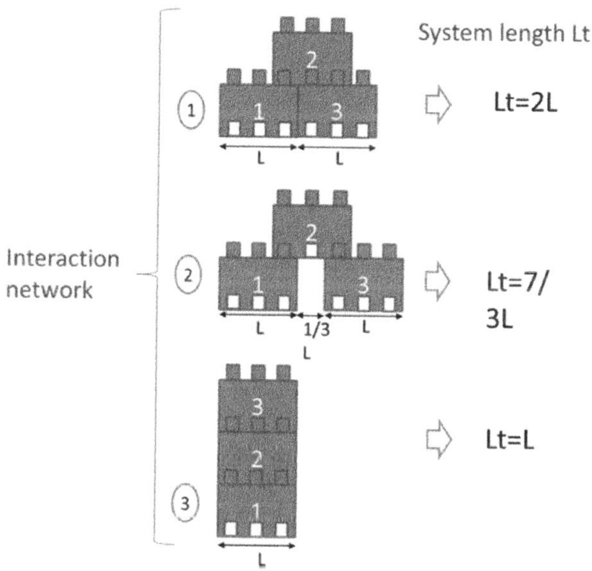

Figure 1.11. *Three possible assembly networks*

– **Time-dependent** properties: properties whose evolution is predictable in time. For example, "every year or every 20,000 km, the system must be serviced".

– Properties **in response to a stimulus**: properties that arise from an external event. We define how the system should respond to an event. For example, "*when the user enters a natural language query, the search engine should provide a list of websites ranked by relevance*".

– **Subjective** properties: properties whose effects depends on the appreciation of an external observer. This is typically the kind of property that has no place in a specification. For example, "*the system must be nice*" or "*the system must be comfortable*".

1.5. The systemic method

The systemic approach is a highly iterative process. The study of a system with its external environment as well as the study of the interior of the system is done in an incremental way and with many loops. We can structure the different steps of this systems approach based on the method resulting from the work on the systemic procedure of Donnadieu and Karsky (2002) and is represented in Figure 1.12.

At the top of the process, we have the **observable**, which is the phenomenon we wish to understand. From there, we gather all the information about the phenomenon. In order to structure this search for information, we will rely on a method that is **systemic triangulation**. With the help of this information and the method, we will proceed to a key step which is the **systemic exploration** that will allow us to organize the different knowledge.

The ordered representation of knowledge is the entry point for **qualitative modeling** of the system. This modeling allows us to represent the static and dynamic structure of the system. We also construct the different causal relationships between the variables characterizing the system with its external environment. Each of these variables must be quantified in order to proceed to the next step.

The **dynamic modeling** of the system consists of establishing the quantitative laws, in other words the equations, which link the various variables. This step will allow us to formalize the system model. The method used is based on the system dynamics of Forrester (1968) and taken up by Aracil (1984). The advantage of this dynamic modeling method is that it represents all types of systems, for example, social, economic and of course technological systems. The procedure will remain the same, but the quantitative relationships will depend on the domain of knowledge.

Dynamic modeling introduces the notion of time in the relationships between variables.

This allows us to move on to the behavioral **simulation** stage. This simulation allows us to test scenarios to improve our understanding of the system and its external environment.

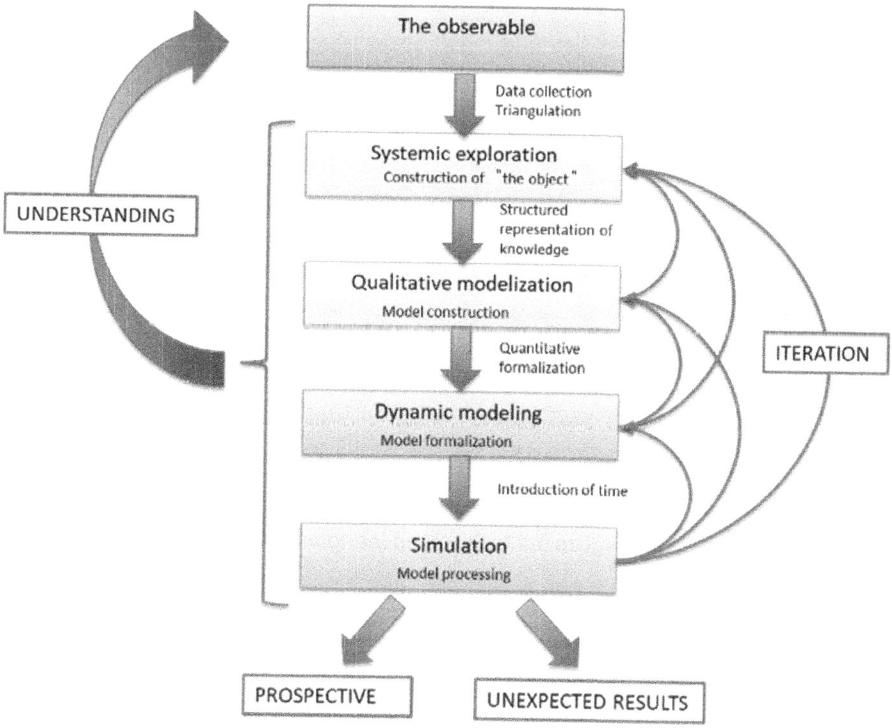

Figure 1.12. *Systemic approach (Donnadieu and Karsky 2002)*

1.5.1. *Systemic exploration*

The objective of systemic exploration is to define the limits of the system. That is, the boundary between the system and its external environment. The exploration will also aim to identify the interactions between the system and this external environment. From there, it will then be possible to look at the inside of the system with its internal interaction network and how it connects and behaves with its external environment. The method we will present to make this exploration is the systemic triangulation proposed by Le Moigne (1983). In Figure 1.13, we have a representation of the system with the three axes to be explored which are the three angles of a triangle. Following each axis, we will study the system according to three visions: operational, functional and organic.

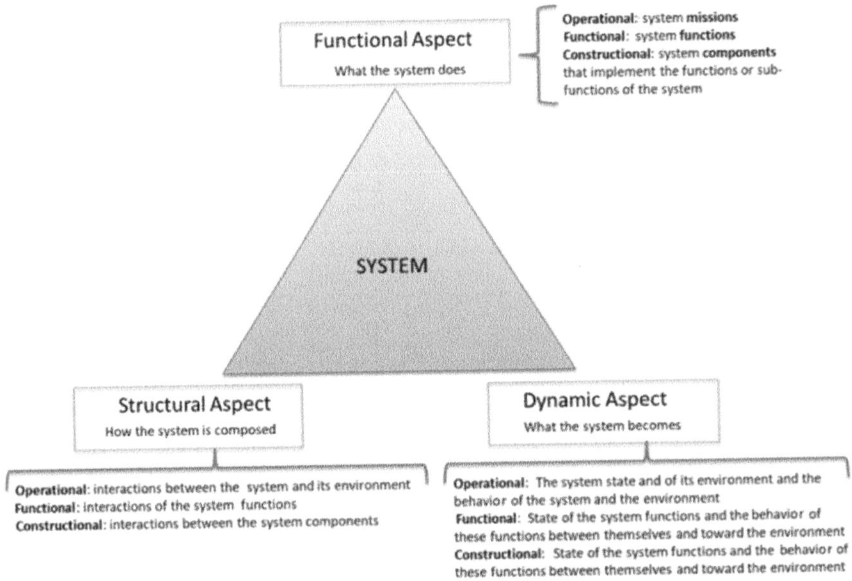

Figure 1.13. *Systemic triangulation*

We will see the meaning of these three points of view in the architecture part of systems engineering.

Functional aspect: on top we have the functional aspect of the system. This aspect allows us to explore the system's purpose by identifying the system's missions and what it must do to achieve them. That is, what functions it should have. The exploration can continue by identifying the physical, computer and human components that will participate in the realization of these functions.

Structural aspect: bottom left, we have the structural aspect of the system where we will look at the complex network of internal and external interactions of the system and its external environment. We can successively and in an iterative way explore the interactions of the system and the external environment. We will also be able to study the internal interactions of the system functions with several levels of decomposition, as well as the interactions between the functions and the external environment. Finally, we will be able to study the organic structure of the system with the components that constitute it, as well as their internal and external interactions with the environment.

Dynamic aspect: bottom right, we have the dynamic aspect of the system. In the literature, we also speak of history or genetics (Donnadieu and Karsky 2002), but we prefer the term dynamics to make the link with the rest of the book easier. The systemic exploration of this aspect consists of studying the system's event dynamics, through the states, as well as its behavior with scenarios. We will be able to study in an iterative way, from an operational point of view, the operational contexts of the system, which are the states of the environment, the functional modes of the system, which are the functional states, and from an organic point of view, the configurations of the system, which are the states of the components of the system. The changes of state are achieved by means of external or internal events to the system, which is why we speak about state or event-driven dynamics. In addition to this analysis of state changes, we can explore the dynamics of the system's behavior according to the three operational, functional and organic points of view. According to the point of view, we can study:

– the overall behavior of the system in its environment to achieve a mission;

– the behavior of a system function in its environment;

– the functional behavior of a system component in its environment.

1.5.2. *Qualitative modeling*

From the set of data resulting from the analysis of the system, we will be able to build a model. This qualitative model will be descriptive and will have a dual purpose. The first is to be able to share understanding of the system with other stakeholders. The second is preparation for building a quantitative model for simulation purposes. The qualitative model takes the form of more or less formalized diagrams.

If we take the three aspects of systemic triangulation applied to the three points of view of the system, we have three families of diagrams, namely tree diagrams, block diagrams and dynamic diagrams. For the functional aspect, we will mainly use tree diagrams because we model the interactions in the structural aspect with block diagrams. Figure 1.14 presents a tree structure with several hierarchical levels. Each box of Figure 1.14 can be replaced by:

– an external system or stakeholder to represent the external environment of the system;

– a mission that can be broken down into elementary missions;

– a function that can be broken down into sub-functions;

– a system with its decomposition into subsystems and equipment;

– a component of the system with its decomposition into sub-components.

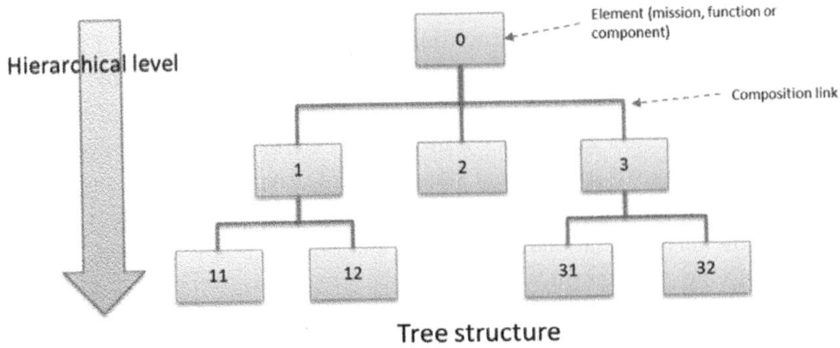

Figure 1.14. *The tree model*

For the structural aspect, we will feed the system model with static views of the block diagram type. Figure 1.15 shows two categories of block diagrams. On the left, we have a box/arrow block diagram that allows us to quickly model matter, energy and information relationships between:

– the system and the external environment;

– the functions between them and with the environment;

– the components between themselves and with the environment.

This type of modeling has the advantage of maintaining a global point of view. The block diagram on the right adds the concept of a port on the elements that allows them to be independent and interchangeable with respect to their respective environments. Ports are interfaces that characterize the expected input and output of the element, which can be either the system, an external system, a function or a component.

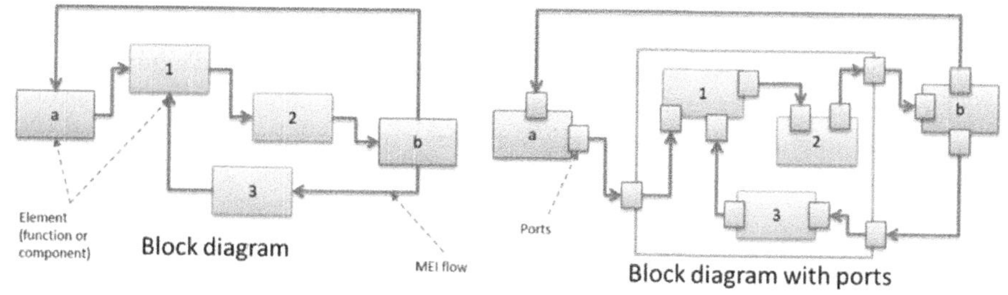

Figure 1.15. *The structural model*

The dynamic aspect of the systemic triangulation will allow us to enrich the model with the evolution in time of the system and its environment. We will model two types of dynamics, namely event-driven dynamics with the evolution of states and behavioral dynamics. Figure 1.16 presents on the left a state machine with events that make it possible to go from one state to another. States can be replaced by operational contexts, functional modes or technical system configurations. Events are conditions that allow a change of state. On the right, we have a sequence diagram modeling the behavior of the system in the form of scenarios. Each element can represent an external system, the system, a function or a component depending on the systemic level we model. Then, the vertical dotted line represents time. The arrows are the exchanges of matter, energy and information between the elements. In order to make the connection with quantitative modeling, it is necessary to introduce a new structural diagram, the causal diagram, which allows the variables of the system and its external environment to appear.

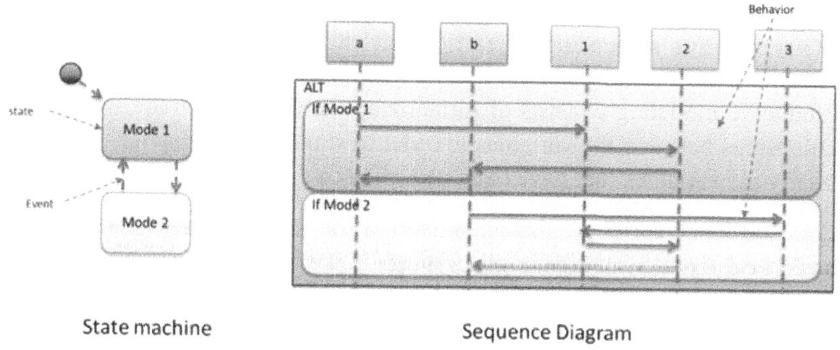

Figure 1.16. *The dynamic discrete event model*

Figure 1.17 shows an example of a causal diagram which is similar to a block diagram of the influence diagram type. The elements on this diagram are the variables of the system, external systems, functions or components. The arrows represent the interactions between the variables, i.e. the influences. The example models a water tank that is to be filled to a desired water volume setpoint.

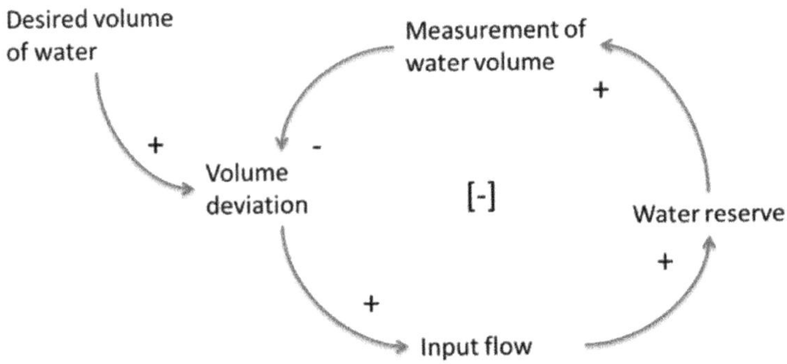

Figure 1.17. *Causal diagram*

It is a negative feedback loop with a desired volume of water representing the goal. Effectively, because the number of negative influences in the loop is odd, it results in a stabilizing effect. The volume difference is the decision that results in an action that is the input flow to the tank. This flow rate is used to refill the water supply, the volume of which is measured in order to compare it with the target and thus determine the deviation to modify the action.

1.5.3. *Quantitative modeling*

Quantitative modeling consists of formalizing the causal diagram by identifying the relationships between the variables in order to simulate its behavior. For this, we will classify the variables that have been identified to distinguish:

– stock or level variables that characterize the states of a system (function, component or external system) at a given time;

– the flow variables that feed or empty the stock variables;

– the constants;

– auxiliary variables that are compositions of stock, flow and constant information or other auxiliary variables. These variables model the nonlinearities of the system.

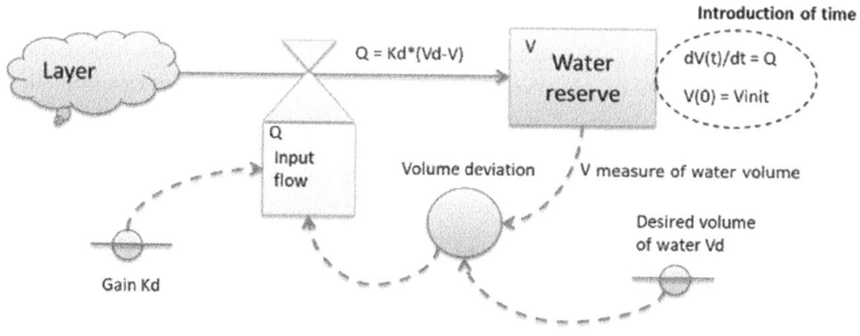

Figure 1.18. *Formal diagram of stocks and flows*

The relationships between these variables materialize information links for constants and auxiliaries or material links for stocks or flow variables. Figure 1.18 shows a stock and flow diagram, also called a Forrester diagram, applied to an example of a water tank. Among the variables in the causal diagram, we have:

– the water reserve which has become a state variable of the system, with an initial value V(0);

– the input flow has become a flow variable that feeds the "water reserve" stock variable;

– the desired volume is a constant; it is the set point to reach;

– the volume difference is an auxiliary variable that is a composition of the water supply information (formerly the water volume measurement variable) and the "desired water volume" setpoint;

– the gain Kd is a constant that has been added.

1.5.4. *The simulation*

To be able to simulate the behavior of a system, it is necessary to add the time variable. Time appears in the stock variables which will be formalized by differential equations. The objective of the simulation is to gain a better understanding of the system. It will allow us to test scenarios by changing, for example, the initial values of the stock variables. However, we must not forget that this simulation is only suitable if we have knowledge of the field it represents, i.e. the corresponding trade or trades. The simulation is built by transforming the formal stock and flow diagram into an equation. As the simulation is finite and in discrete time, we must define the length, in the sense of duration, of the simulation as well as its integration step. Figure 1.19 shows the simulation results of the water tank that we modeled with a Forrester diagram. The outputs of the simulation are time curves representing the evolution of observed variables. We can see in our example that two variables have been observed, namely the flow variable which is the inlet flow of the tank and the stock variable which is the volume of the tank.

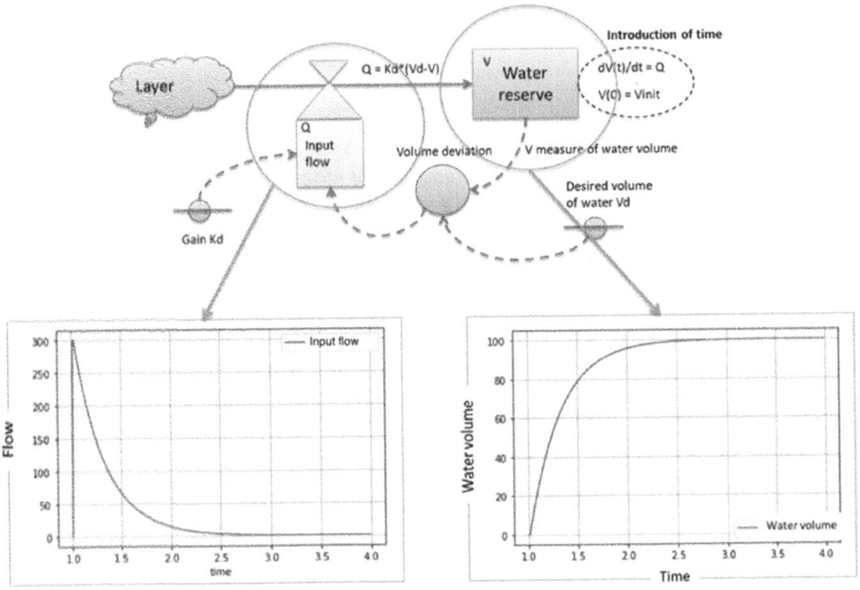

Figure 1.19. *Simulation result*

1.6. How to understand the complexity of a system

We can understand a complex system according to **four themes** (AFIS 2012):

– the system is in an **environment**: the black box point of view;

– the system has an **architecture**: the structural white box point of view;

– the system has a **dynamic**: the event-driven and behavioral white box point of view;

– the system is **controlled**: analysis of feedback loops.

1.6.1. *Theme 1: the system is in an environment*

The system is first seen as a **black box** (external view) in an environment. Knowledge of the environment and its properties is essential. Indeed, it is the environment that defines the **purpose** of the system, which is its raison d'être or its goal. This goal is expressed in the form of a mission that identifies the expected services. The system is in interaction with this environment. It transforms inputs of matter, energy and information into outputs of matter, energy and information in order to meet the services expected by the environment. It also respects the constraints imposed by the latter.

This theme of black box and environment implies **two properties**. The system must be **open** in order to exchange flows. The system must have a **finite perimeter** so that we can distinguish between the inside and outside of the system. Boirel (1988) represents the system and its environment as concentric environments.

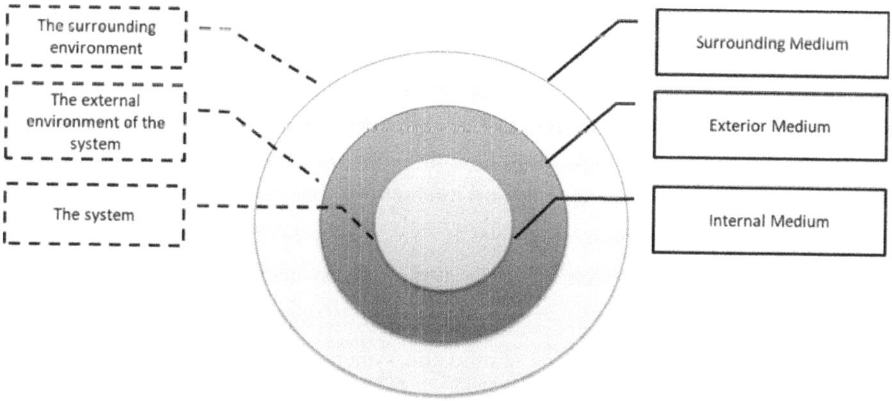

Figure 1.20. *The environments*

In Figure 1.20, we have the internal environment which gathers all the internal interactions of the system. Then, we have the external environment, which is the environment with which the system interacts, i.e. the intervention space. Finally, we have the encompassing environment, which is the environment in the broad sense that integrates the system and the external environment.

We can complete this "medium" approach with the formal definition of the environment proposed by Krob (2014). The environment E(S) of a system is the integration of the system S and an external system Ext(S) (see Figure 1.21). We find in this definition the three environments of Boirel.

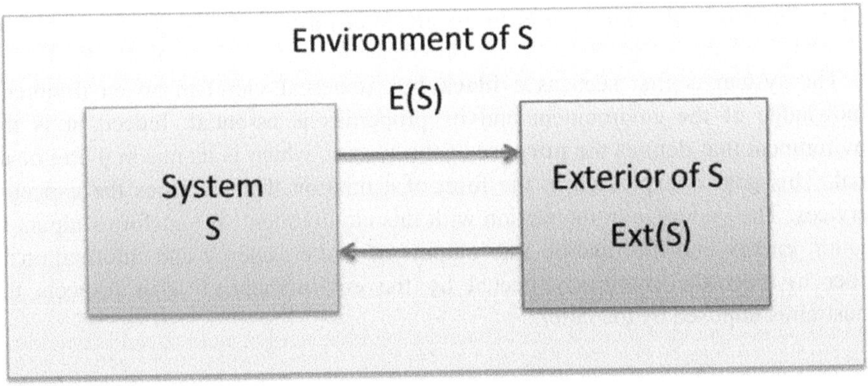

Figure 1.21. *Formal environment (Krob 2014)*

1.6.2. *Theme 2: the system is structured*

After defining the boundaries with the external environment, the system can be seen as a **white box**. That is, we open the box to see what is inside. This is an internal view. The system is composed of elements that interact with each other and with the environment. We will study how these interactions are regulated to achieve the system's purpose. The elements and interactions that make up the system can be studied from:

– a functional angle: what does the system do to achieve the purpose, i.e. its goal;

– a constructional angle: how should the system implement the functions that enable it to achieve the purpose.

The system can be seen as a functional or organic network that exchanges flows of matter, energy and information. In this network, the elements can have different roles. There are elements that are reservoirs, they will store matter, energy or information. Other elements will transform matter, energy and information. There are also elements that are there to circulate matter, energy and information. Finally, there are elements that have a regulatory role and serve to control interactions.

1.6.3. *Theme 3: the system evolves over time and has a dynamic*

The system evolves over time and adapts to its environment. This adaptation is necessary because the environment also changes over time. Indeed, to study a system, we must analyze all of its life situations, which we will call operational contexts. An operational context is a state of the environment during which the external systems that define it are in a certain state or set of possible states. To cover these contexts in terms of uses and constraints, the system itself must have functional and organic states that we will call operating modes and technical configurations. Depending on these states, the system adopts a dynamic behavior that results in the activation of functional chains. This dynamic behavior is the result of the integration of the static point of view with the interacting elements and the event-driven dynamic point of view with the states of the system and the external environment. In system dynamics, behavior can be characterized by identifying stock and flow variables and their relationships (Forrester 1968). The stock variables, also called level or state variables, describe the situation of the system at any time. The flow variables, also called rate variables, reflect the speed at which stocks change. In system dynamics, the modeling of variables concerns the system and its external environment.

1.6.4. *Theme 4: the system is controlled*

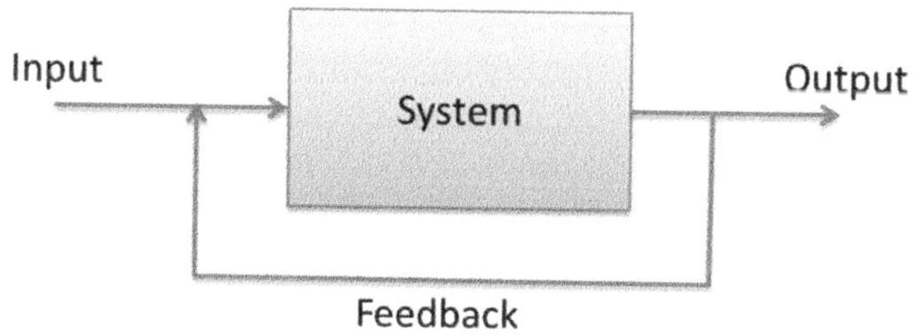

Figure 1.22. *The feedback loop*

To be able to adapt and achieve its missions, the system must be controlled or regulated to be free of (Pollet 2016), for example:

– environmental disturbances;

– human errors;

– system reliability defects such as software bugs.

The concept of regulation is based on feedback loops (see Figure 1.22). We can return to the property of an open system that transforms incoming flows into outgoing flows. There is no link between incoming and outgoing flows. In an open system, outflows have no influence on inflows. The system is not informed of its own functioning and its past has no influence on its future. A system with a feedback loop is influenced by its past behavior, and the outgoing flows are fed back as information by a control function to the incoming flows (Forrester 1968; Donnadieu and Karsky 2002). Figure 1.23 illustrates the feedback loop mechanism. We have a measurement function that acquires the system's output flows, which can be matter, energy or information and transforms them into information. Then, we have a comparator that makes the difference between the goal to be reached, which we also call the setpoint, and the measurement that has just been made. Finally, this deviation is entered into a control function that will calculate the new inputs of our system to be controlled.

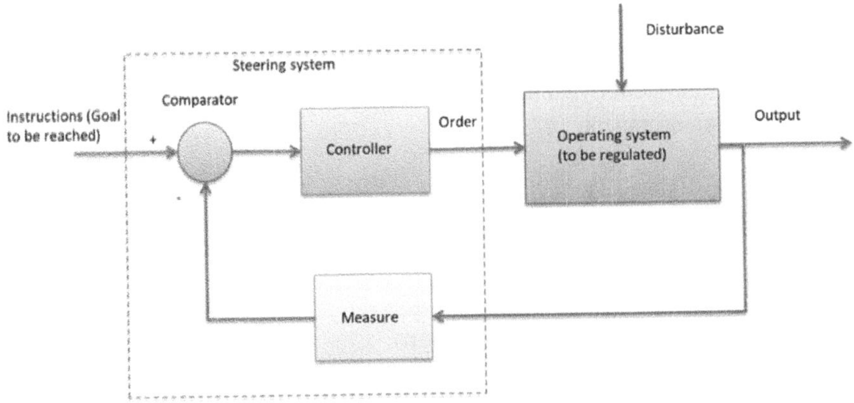

Figure 1.23. *Example of a feedback loop or cybernetic loop*

Le Moigne (1999) presents the nine levels of organization of a complex system model. Figures 1.24–1.28 present the modeling of the nine levels of a system's complexity.

Figure 1.24. *Modeling of level 1 and 2 complexity systems*

The level 1 system has a perimeter and can be distinguished from its environment. In the level 2 system, we see the first functions appear.

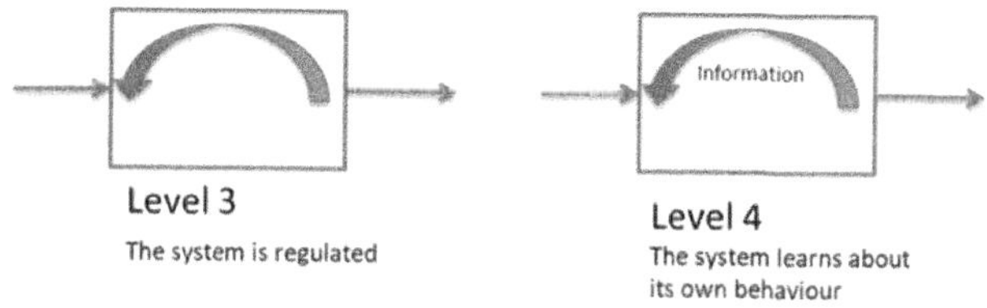

Figure 1.25. *Modeling of level 3 and 4 complexity systems*

Regulation appears on level 3 and 4 systems. On the level 3 system, regulation is assumed as it is not observable. On the level 4 system, the regulation is done through observations of the system.

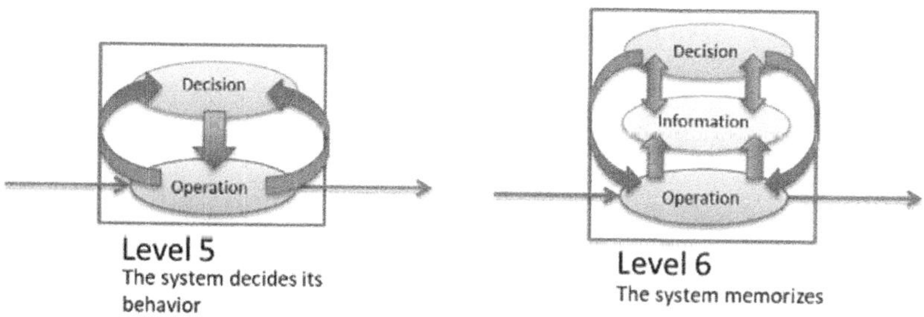

Figure 1.26. *Modeling of level 5 and 6 complexity systems*

We can clearly see the enrichment of the decision subsystem that appears from level 5.

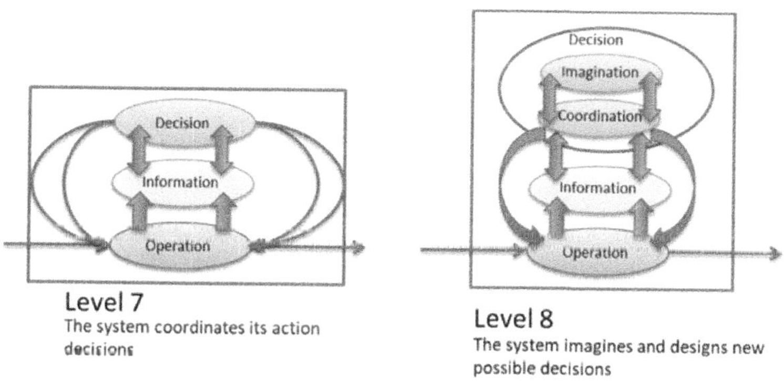

Figure 1.27. *Modeling of level 7 and 8 complexity systems*

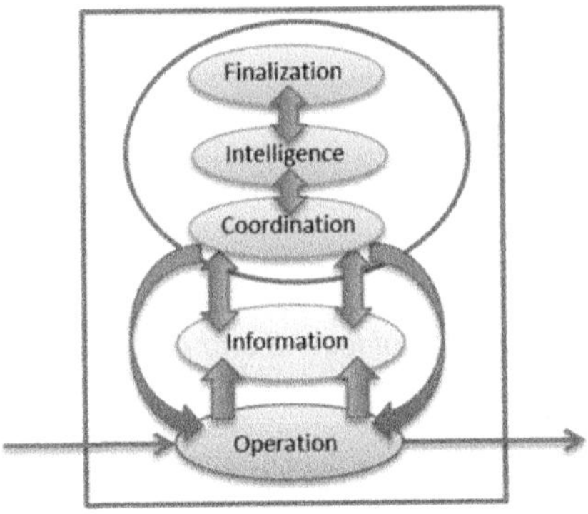

Figure 1.28. *Level 9 system modeling*

The table in Figure 1.29 summarizes and describes each level of a system's complexity.

Level	Type	Description
Level 1	The system is identifiable	The system has a limit. It differentiates from its external environment
Level 2	The system is active: It acts	The system transforms the input flows into output flows
Level 3	The system is regulated	The system has an internal regulation presumed and non-observable
Level 4	The system learns about its own behavior	The system is regulated and transforms the flows to information for its regulation
Level 5	The system decides its behavior	The system generates information. It is able to regulate itself
Level 6	The system memorizes	The system memorises information. There is a structure of the type operation, information, memorisation and decision
Level 7	The system coordinates its action decisions	The system coordinates its actions
Level 8	The system imagines and designs new possible decisions	The system is able to elaborate new actions. The decision sub-system is divided into a sub-system of imagination and a sub-system of coordination
Level 9	The system terminates	The system is able to decide its own purpose. The decision system decomposes into a sub-system of finalization, a sub-system of intelligence and a sub-system of coordination

Figure 1.29. *Complex nine-level system*

We will present in more detail levels 5 and 6 which are well adapted to the modeling of most technological systems. We find them regularly in the literature (Pollet 2007; AFIS 2012). The level 5 system of complexity on Le Moigne's scale is the first to distinguish between system control and operation. This system is known as the cybernetic system (Wiener 1948). In Figure 1.30, we have an abstract view of a piloted system made of two subsystems. First, we have the operating subsystem which is the part of the system that contributes to the value by meeting the goals of the user environment. The second part is the control subsystem which contributes to the regulation of the system and allows it to adapt to the disturbances of the environment of use. This first model of a complex system can be used for the architecture of the system and on the three points of view, namely operational, functional and organic, of the system.

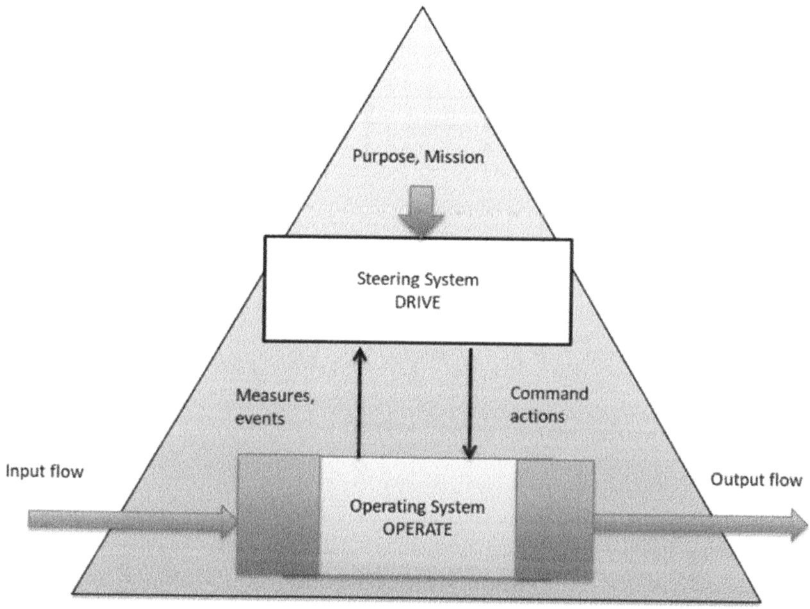

Figure 1.30. *Controlled system*

For example, the implementation of this model's:

– operational view can be realized by identifying the driving flows for the control of the system and the operating flows in input or output for the value contribution of the system;

– functional view can be realized by identifying the functions that drive or regulate the system and the operation functions that bring value to the system by fulfilling the missions;

– organic view can be achieved by identifying the components that drive the system and the operating components that participate in the value of the system by performing the functions that participate in the missions.

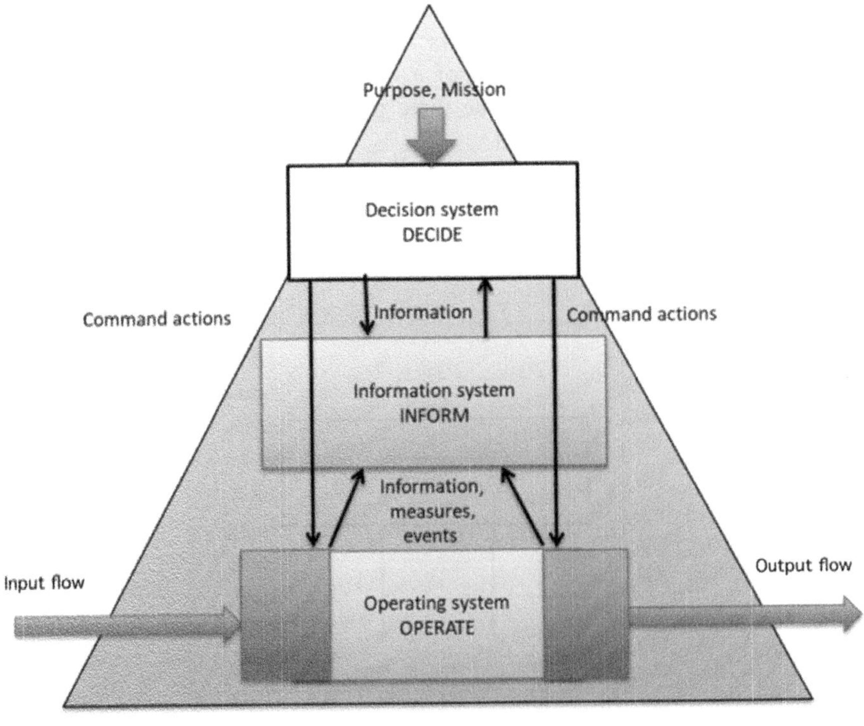

Figure 1.31. *OID model*

Figure 1.31 shows a model in canonical OID form. It models a level 6 complex system out of the nine levels of complexity presented in Le Moigne (1999). This system model is the first one that we can qualify as intelligent, thanks to its capacity of memorization which is a prerequisite. The system has an architecture based on three subsystems:

– the operating subsystem which, as in the previous model, contributes to the value of the system by enabling the user environment to achieve its purpose;

– the information subsystem which enables memorizing and communicating the measures and actions of the system and the information of the environment;

– the decision subsystem that regulates and adapts the behavior of the system with this information.

Like the first model, the canonical OID form can be used to architect a system on these three points of view. For example, it can be representative of a complex system with preponderant software.

1.7. Conclusion on systems theory

The systemic approach is based on the principle of **totality** where the system we call the "whole" is more than the sum of its parts, i.e. its constituents. This highlights the property of non-summability of complex systems.

To understand this complexity, we will try to model the network of internal and external interactions of the system by showing the feedback loops. We have defined the notion of system which is a starting point in the systemic approach. What we must remember is that a system is a set of interacting elements, that it has a purpose and that it is in a given environment.

Among the systems, we have defined the complexity and how to make sense of it. This complexity is defined by the size of the interaction network with cause and effect links and the presence of feedback loops. In addition, there are the key factors, delays and threshold effects.

A complex system can be modeled from a static and dynamic perspective. The static model allows us to highlight the system's internal and external interaction network. The dynamic model builds or reconstructs the behavior of the system and the evolution of its internal states. The modeling also focuses on the system's control by analyzing the structure and dynamics of the feedback loops.

PART 2

Systems and Requirements Engineering

2

Introduction to Systems Engineering

2.1. The system meets needs

2.1.1. *Definition of a need*

In general, the system must meet a **purpose** and it is from this purpose that the **need** is derived.

DEFINITION.– A need is a necessity, desire or expectation felt by a stakeholder. The needs are formalized into stakeholder requirements (AFIS 2012).

The definition of a system states that it must meet a certain **purpose**. The purpose of a system is the main goal to which it must respond. To achieve this **goal**, the system performs **missions** for an environment of users; these users determine the missions of the system. These missions are put in the form of **operational scenarios** and must satisfy quantitative **objectives**.

To achieve its missions, the system offers **services** to users and respects the **constraints** of the environment of use. These services and these constraints reflect the **needs** of the environment with respect to the system. In order to be analyzed and managed, these needs will be formalized into **stakeholder requirements**.

For a color version of all the figures in this chapter, see www.iste.co.uk/hygounenc/system.zip.

Systems Engineering,

2.2. Definition of a stakeholder requirement

DEFINITION.– A stakeholder requirement (also called a user requirement) is a condition or capability that a user needs to solve a problem or achieve a goal (Stéphane and Badreau 2014).

A stakeholder requirement is therefore a property of that stakeholder. These properties are formalized and managed in the same way as the system properties.

2.3. How to go from need to system

As an example, in Figure 2.1, on the top left, we have the need: "the desire to fly". On the bottom right, we have the system, an ultralight, which satisfies this need. Note the notion of a closed loop after the transition between the need and the system. This looping is called the **validation** of the system with respect to the user's need. To go from the need to the system, we will implement a systems engineering approach.

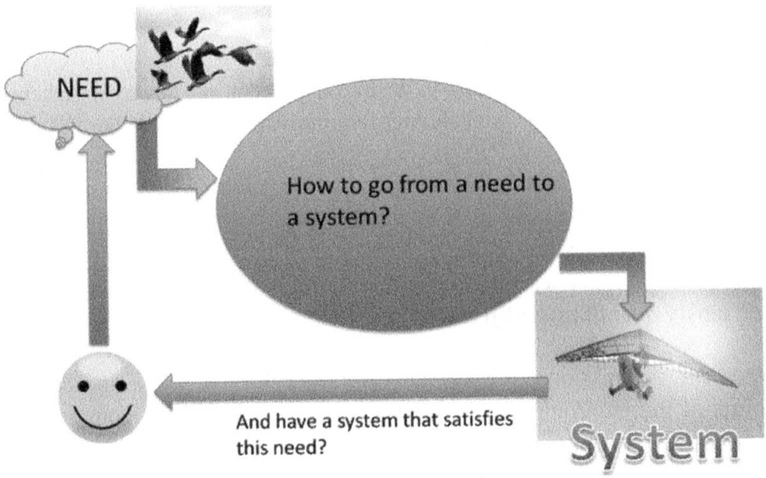

Figure 2.1. *How to go from need to system*

2.4. Definition of systems engineering

Systems engineering is a collaborative and interdisciplinary approach. It is based on science and **experience** and encompasses *the activities* to design, develop, evolve and verify a **system**. Indeed, the objective of systems engineering is to allow engineers (in electronics, mechanics, computer science, etc.), cross-functional engineers (safety, security, quality, etc.), managers and customer representatives to share the **same vision of the system to be built**.

Systems engineering is the application of the systems approach to business engineering (see Figure 1.6), which Thomé (1993) calls systems engineering.

Systems engineering concerns the **entire life cycle** of the system by focusing on the system under study, as well as on all the contributing systems (see Figure 2.2).

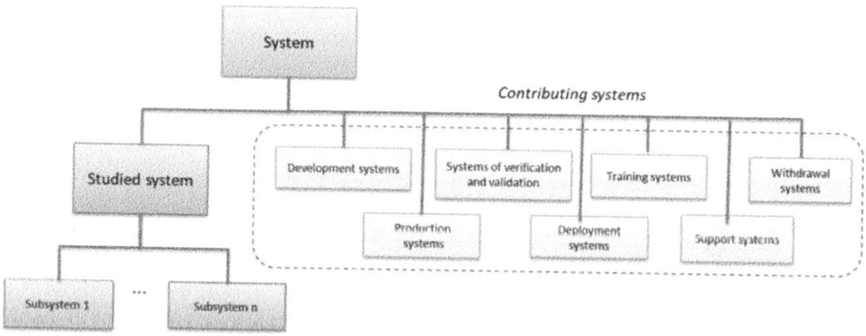

Figure 2.2. *Systems of interest with contributing systems, (IEEE-1220 found in IEEE (2017))*

Systems engineering is a highly **iterative** approach. Indeed, if we base ourselves on the V-shaped development cycle, the activities seem sequential: analysis of the needs, then provision of a specification, then design, then realization, afterwards integration, then verification and acceptance of the final product. This apparent linearity is attractive for project planning, but in systems engineering, when the architect delivers a version of their specification, they will have already done several iterations on their design. For this, systems engineering allows the engineer to **work in the blur AND to know where they stand**!

And of course, the goal of systems engineering is to provide **a solution** that is:

– optimized for identified needs;

– environmentally acceptable.

The term "identified need" is important because we will see that a client rarely formalizes all their needs, hence the need to go and find them.

2.5. Iterative systems engineering process

2.5.1. *Prescription and system construction*

To go from the need to the system, we will implement a systems engineering approach, composed of four iterative technical processes and two supporting processes (AFIS 2012).

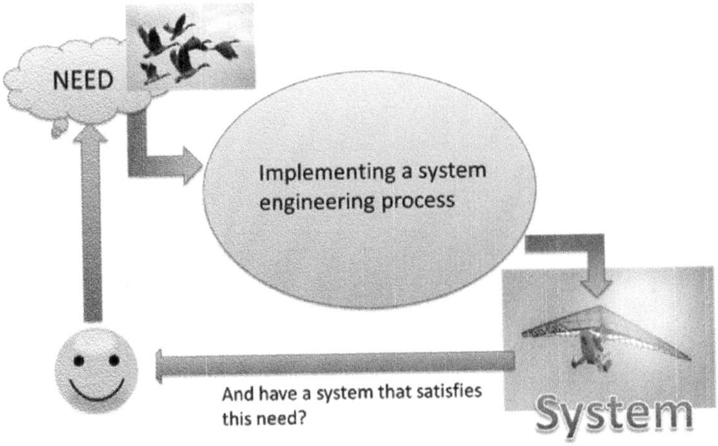

Figure 2.3. *The systems engineering process*

On the upper part of Figure 2.4, we are in the world of system prescription; the system is defined by its requirements in response to needs. In the lower part, we are in the world of system construction. The system is built from a functional and organic perspective in response to its needs and requirements.

Figure 2.5 presents an iteration of these processes in a sequential form with activity verification loops and validations between two activity steps. We can also distinguish the passage from the need to the solution by having as a boundary the formatting of the problem to be solved.

The **needs analysis** process enables all of the expectations to be collected and formalized in terms of the uses and constraints of the stakeholders of the problem domain and the solution domain. From the uses, the missions and the operational contexts of the system are identified.

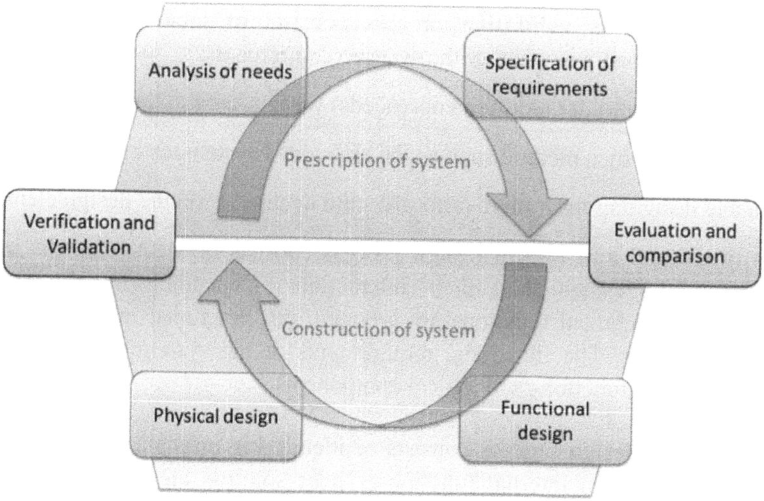

Figure 2.4. *Iterative processes of systems engineering*

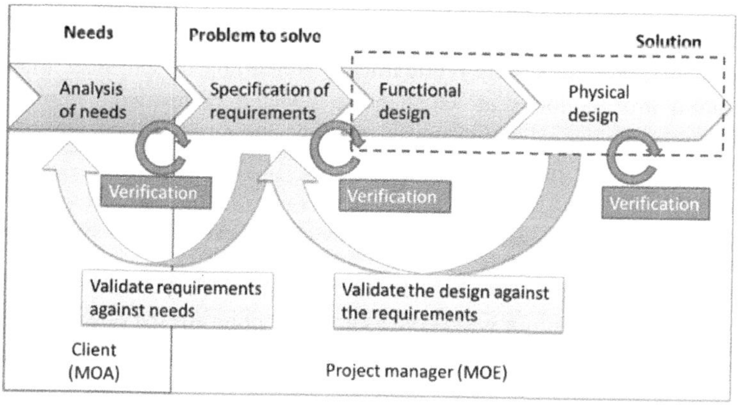

Figure 2.5. *Systems engineering process*

The objective of the **requirements specification** process is to formalize all the needs of the stakeholders into functional and non-functional properties of the integrated system. Functional properties are grouped according to the system to meet the missions, and operational contexts are covered by the functional modes of the system.

The **requirements validation** process consists of looking at whether the problem is well posed, i.e. whether the system requirements:

– cover all stakeholder requirements (needs) well;

– are verifiable by a method such as the embedded system test;

– enable designing the system (allocable and declinable on the design elements).

The **functional and logical design** process consists of identifying the internal functions of the system and their interactions. Depending on the system's complexity, the technical functions are grouped into functional modules that form logical components. The functional requirements of the system are allocated and derived on the sub-functions and logical components.

The **physical design** process consists of identifying possible solution elements and allocating them to internal functions. These solution elements are structured by identifying the physical links between them and the environment. The non-functional requirements of the system are allocated to the different physical components.

The **design validation** process consists of looking at whether the functional or logical design meets the system requirements. Regarding the physical architecture, a verification process looks at whether the selected components implement the sub-functions of the logical architecture.

2.5.2. *Needs analysis*

The objective of the needs analysis is to formalize a repository of stakeholder requirements. This repository will serve as an entry point for the development of the system by posing the problem from the user's point of view, i.e. the project owner.

It will be completed by the requirements of the stakeholders interested in the realization of the system which will bring their constraints on the life cycle of the system (development, realization, commissioning, maintenance, training, withdrawal from service).

This analysis is conducted by building an operational architecture of the system to be made, which is seen as a black box in its environment. To cover all the needs, we study the direct and indirect environments.

The system will be modeled from the following angles:

– static (the missions of the system and the interactions with the environment);

– dynamic (the life cycle and operational contexts of the system);

– behavioral (the operational scenarios which are the mission profiles with the sequence of services).

The properties (expected services and environment constraints) will be formalized into stakeholder requirements so that they can be managed, declined on the system requirements and validated on the final product.

2.5.3. *Requirements specification*

The objective of the requirements specification of the system is to formalize a technical repository that poses the problem to be solved from the perspective of the project manager.

The system requirements must meet the needs formalized by the users while taking into account the constraints of the stakeholders involved in the realization. There are two kinds of requirements: functional and non-functional.

This analysis is conducted by taking the operational architecture and initializing the functional architecture. The analysis of the missions and the operational scenarios will allow for the transformation of the services into system functions that will introduce the technical principles retained or imposed on the project manager.

The dynamic analysis will identify all the functional modes covering the operational contexts and allowing the activation of the functions.

Behavioral analysis will allow us to reconstruct the missions through the functions of the system.

From this analysis, the functional and non-functional requirements of the system will be formalized. They will prescribe the set of properties that we will find on the integrated system.

2.5.4. *Functional and logical design*

The objective of functional design is to propose an architecture of sub-functions that enable the implementation of the system functions and that can be allocated to hardware, software and human components.

This design activity is carried out from the three angles, namely static, dynamic and behavioral, with the parallel development and refinement of functional and performance requirements.

Static functional design consists of decomposing system functions into sub-functions until they can be allocated to constituents.

In order to facilitate this allocation, the sub-functions can be organized into logical components. The interactions between the sub-functions and the environment are defined.

Dynamic functional design consists of refining the functional modes of the system by identifying, if necessary, the sub-modes and the sub-functions that are activated.

This activity is completed by the identification of the steering sub-functions that are added to the functional decomposition.

Functional behavioral design is the integration of the static and dynamic vision. Following the evolution of the functional modes and the static functional architecture, the functional behaviors reconstruct the functions and missions of the system by activating functional chains of the sub-function network.

2.5.5. *Physical design*

The objective of physical design is to propose an architecture of hardware, software and human components, executing the sub-functions of the functional architecture, to address the missions and the constraints of the operational architecture and to allow its realization by the industrial organization.

System requirements will be allocated regarding this architecture in order to select candidate solutions. To enable its realization, one of the outputs of the physical design will be to specify the physical architecture with the requirements defined for the subsystems that have been selected.

This physical architecture is obtained using a top-down and bottom-up approach. The top-down approach uses the outputs of the operational and functional architectures, and the bottom-up approach relies on the know-how of the project manager and the state of the art, in addition to the company's innovation strategy, which is part of risk management.

As with functional architecture, this physical design activity is carried out from three perspectives: static, dynamic and behavioral.

The static physical design consists of identifying the components of the solution that can realize the sub-functions or logical components of the functional architecture. Physical components can be grouped into subsystems, according to architectural principles and in order to respect operational constraints. Physical interfaces must be identified.

The dynamic design consists of identifying the technical configurations of the system translating the functional modes through these components. Each configuration or sub-configuration activates a set of components that perform system functions.

The physical behavioral design reconstructs operational scenarios with the system components.

During these activities, the system requirements are allocated to the components and specified when the selection has been made. This repository of specified requirements will be used to implement the system.

2.5.6. *Verification and validation*

The verification and validation (V&V) activities are numerous during the development of a system. It is not only limited to the integration phase of the subsystems.

However, V&V activities still meet the same objectives, namely:

– For validation: Does my system meet the needs?

– For verification: Does my system meet its requirements?

These two questions can be adapted to the different stages of development (see Figure I.1 of the AFIS diagram on system development).

2.5.7. *Evaluation and comparison*

The evaluation and comparison process is a support process for the requirements definition, design and V&V activities.

For example, when specifying requirements, several functional choices may be possible and must be evaluated and compared.

Similarly, during requirements validation, needs and requirements may be subject to trade-offs that must be evaluated and compared.

Functional and physical designs can result in different architectures that meet the needs and requirements. These candidate architectures will be evaluated and compared.

For example, one approach is to identify the value and risk criteria common to all these solutions. Each architecture will be assigned a value score and a risk score. The results will be displayed on an Eisenhower matrix-type diagram in order to make a decision.

2.5.8. *The requirements cycle*

The requirements (in the broadest sense) are handled by a form of engineering, which is integrated with systems engineering; this is **requirements engineering**.

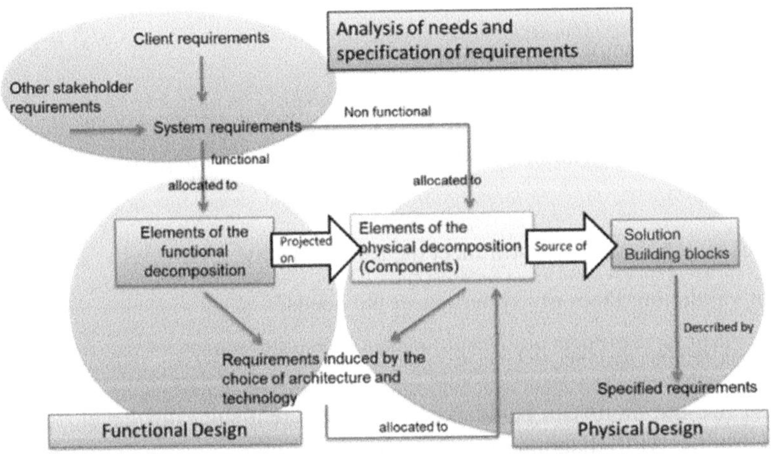

Figure 2.6. *Requirements and systems engineering processes according to the EIA-632 standard (ANSI 1999)*

Figure 2.6 makes the link between the requirements and the systems engineering processes. This principle of transformation and allocation of requirements is taken from the EIA-632 standard (ANSI 1999).

Customer requirements and other stakeholder requirements are the formalization of needs into stakeholder requirements. The system requirements are the translation of this need under a technical angle and put in a specification, which can be a document or a database.

Then, we have the allocation of system requirements, namely "functional" and "non-functional", on the design elements. We also have the appearance of requirements induced by design choices.

And finally, we have, after the selection of the solution, the definition of the system through the specified requirements of its components.

2.6. System architecture

In Krob (2017), **system architecture** is the discipline that synthesizes the methods and tools that allow the **exhaustive and coherent modeling of** a **system** in order to **effectively manage** its implementation throughout its life cycle (design, deployment, maintenance, etc., withdrawal of service).

This exhaustive and coherent modeling of the system is achieved according to three operational, functional and organic dimensions (Krob 2009) (see Figure 2.7).

Views	Modeling parameters	Answer the question	Type of analysis
Operational	External interaction of the system	Why?	Analysis of the environment and the system (black box)
Functional	Abstract functions of the system	What?	Abstract analysis of the system's interior (white box)
Constructional	Concrete components of the system	How?	Concrete analysis of the system's interior (white box)

Figure 2.7. *Architectural visions*

The operational vision allows us to answer the question of WHY, i.e. What is the purpose of the system, what are its missions?

We will identify the system's environment as well as the interactions with it in order to fulfill all of its missions. We will also analyze all the constraints imposed by the environment.

The functional vision allows us to answer WHAT, i.e. what does the system do to fulfill its missions. To do this, we will study the functions of the system and how they work. The operation of a function is expressed by studying its behavior through its sub-functions.

The constructional vision allows us to answer the question WHY, i.e. what the system is made of, i.e. the BEING of the system.

To do this, we will translate the operation into sub-functions that will be allocated to hardware, software and human components. We will study the interfaces and the constructional functioning of these components between them in order to meet the functions and missions, as well as the constraints of the environment.

The WHAT and the HOW are the two levers available to the system designers to answer the WHY. Indeed, the functions and sub-functions respond to the missions and the components respond to the functions and sub-functions as well as the constraints.

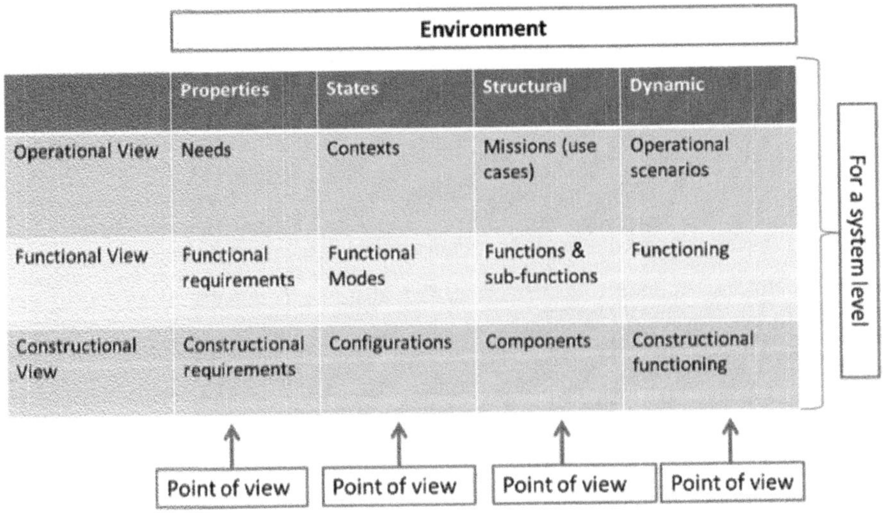

Figure 2.8. *CESAM architectural framing*

For modeling these three architectural visions, we can implement the CESAM architectural framing (Krob 2009, 2017). In Figure 2.8, we have at the top the system's environment composed of external systems and stakeholders that have a direct or indirect influence with the system to be made. The environment is above the grid, as we will find it on all three architectural visions.

The grid represents the system with the three architectural visions and a set of columns. The intersection between a column and a vision forms what is called a viewpoint of the system to be made.

The *first column* is the property column with the system needs and requirements.

The *second column* is the status column with the operational contexts, the functional modes and the technical configurations of the system to be made.

The *third column* is the static elements column with the missions, functions and components of the system.

The *fourth column* is the dynamic elements column with the operational scenarios, the functionings and the constructional scenarios of the system to be made.

2.7. V-cycle

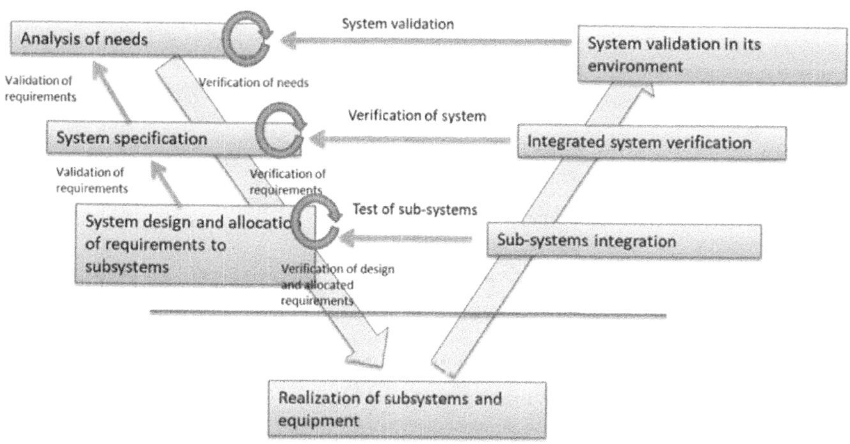

Figure 2.9. *V-cycle*

The V-shaped development cycle is the design process of a system presented in the form of a linear cycle (Figure 2.9). On the way down the cycle, we find the four technical processes of systems engineering with verification and validation steps. At the bottom of the cycle, we have the realization of the system's components. Then, on the upstream part of the cycle, we have the integration of the system with the system's stages of verification with respect to its requirements and validation of the system with respect to the operational needs of the users.

In Figure 2.10, we have a summary of the inputs and outputs of each process covering the entire V-shaped development cycle of a product.

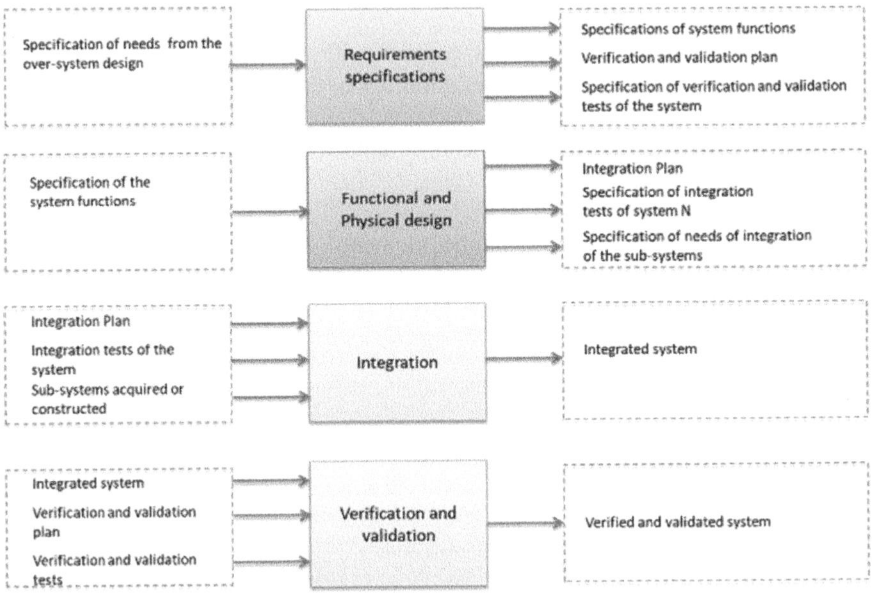

Figure 2.10. *Development of a system*

2.8. W-cycle

The **W-cycle** is an evolution of the V-cycle. It shows upward movements of the cycle during the system's design (see Figure 2.11).

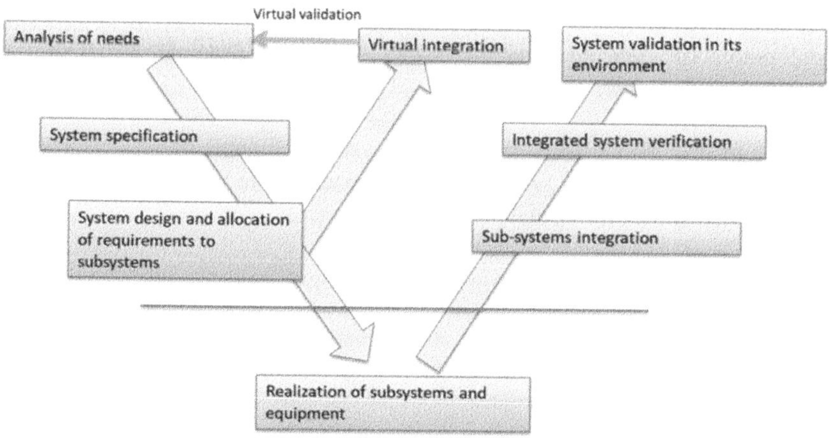

Figure 2.11. *Virtual integration*

These cycle feedbacks are **virtual** system **integration** phases. Indeed, instead of integrating physical elements, we integrate behavioral and multiphysical models of the system.

We also talk about **virtual validation**. For this purpose, behavioral and multiphysical models are integrated with environmental models for the execution of operational scenarios.

2.9. Conclusion on systems engineering

In conclusion, here is what you should remember about this chapter on systems engineering.

Systems engineering is a systems approach applied to business engineering. Systems engineering is an iterative process based on science and experience. It covers the entire development cycle of the system.

It consists of four technical processes which are needs analysis, requirements specification, functional and physical design of the system. It has two supporting processes which are the verification and validation process, applicable on and between the four technical processes, and the evaluation and comparison process.

It can be implemented through the system architecture by applying an architectural framework that carries the three visions of a system, namely operational, functional and organic. This framing models a system according to several points of view such as the static, structural, dynamic, behavioral and properties points of view.

From a project management perspective, systems engineering follows a V-cycle of development. In this book, we introduce an extension to the V-cycle which is the W-cycle, which allows for virtual integration.

3

Introduction to Requirements Engineering

3.1. Definition of requirements engineering

Requirements engineering is a methodical, multi-disciplinary discipline that **collects**, **specifies**, **refines** and **qualifies** requirements related to a system to be positioned in its environment, and **controls the evolution** of these requirements over time. Requirements engineering integrates two types of activities:

– the definition of requirements;

– management of requirements and their evolution.

Requirements engineering is part of the systems engineering activities, as shown in Figure 3.1.

At the beginning, we start with an endpoint that represents a goal that the system must achieve for the users. Stakeholders interested in using the system will express their needs in the form of initial requirements, formalized in an initial specification. This first part is the domain of the project owner.

These initial requirements will be derived, in the "declined" sense, into system requirements and will take into account the requirements of the other stakeholders who are involved in the realization of the system. These system requirements will be formalized in a technical specification. This part is the domain of the project manager and is subject to a compromise and validation by the project owner. These first two steps are in the domain of the system prescription and formalize the problem to be solved.

For a color version of all the figures in this chapter, see www.iste.co.uk/hygounenc/system.zip.

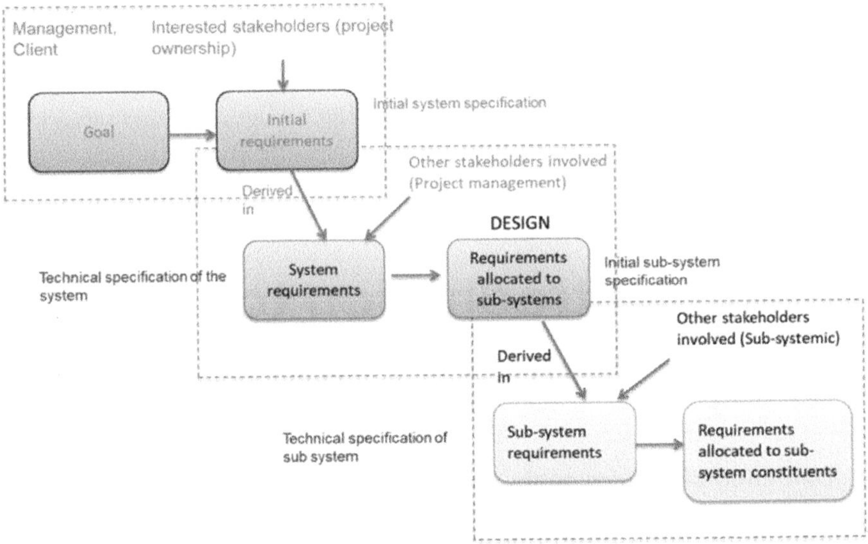

Figure 3.1. *Derivation of requirements*

Through the systems engineering activities of functional and physical design, we will identify the system components. The associated requirements engineering activity consists of allocating the system requirements to the subsystems. This declination can be refined. These allocated requirements are the initial requirements of the subsystems and are formalized in a subsystem specification. The process is thus repeated on each systemic level. This part is in the field of system construction and formalizes the solution to be realized.

To conclude this definition of requirements engineering, the requirements definition activity covers the problem domain and the solution domain. The activity can be broken down as follows:

– formalize the needs of stakeholders;

– formalize system requirements;

– validate system requirements;

– allocate system requirements to subsystems;

– validate the allocated requirements of the subsystems;

– define the requirements of the subsystems.

3.2. The importance and challenges of requirements engineering

According to the Standish Group's CHAOS report (see Figure 3.2), the number one factor in project failure is related to a requirements problem (ProjectSmart 2014).

In fourth place, we have unrealistic requests and in eighth place, the needs that were not. These two sources of failure are linked to a lack of definition of needs. In sixth position, we have as cause of failure the problems of managing the evolution of the requirements.

Causes of success of a project

Project Success Factors	% of Responses
1. User Involvement	15.9%
2. Executive Management Support	13.9%
3. Clear Statement of Requirements	13.0%
4. Proper Planning	9.6%
5. Realistic Expectations	8.2%
6. Smaller Project Milestones	7.7%
7. Competent Staff	7.2%
8. Ownership	5.3%
9. Clear Vision & Objectives	2.9%
10. Hard-Working, Focused Staff	2.4%
Other	13.9%

Causes of failure of a project

Project Impaired Factors	% of Responses
1. Incomplete Requirements	13.1%
2. Lack of User Involvement	12.4%
3. Lack of Resources	10.6%
4. Unrealistic Expectations	9.9%
5. Lack of Executive Support	9.3%
6. Changing Requirements & Specifications	8.7%
7. Lack of Planning	8.1%
8. Didn't Need It Any Longer	7.5%
9. Lack of IT Management	6.2%
10. Technology Illiteracy	4.3%
Other	9.9%

Figure 3.2. *Success/failure of a project (ProjectSmart 2014)*

We can also see that the control of requirements is one of the first three causes of success of a project.

Requirements engineering seeks to address several issues. First of all, with regard to the project owner, requirements engineering will seek **customer satisfaction** by working on the quality of the deliverables and the product.

Indeed, a quality system is a system that satisfies the needs of the user. This is why requirements engineering will improve the identification and coverage of the need.

On the project manager's side, requirements engineering aims to reduce project costs and risks. Indeed, we will see that a mistake on a requirement can be very costly. Requirements engineering improves project control and facilitates exchanges and communication within the teams.

3.3. Problem domain and solution domain

A distinction must be made between the problem domain and the solution domain (see Figure 3.3). The problem domain is covered by the client. They are the manager and integrator of the needs. As a general rule, they formalize their needs in an initial specification.

This specification contains the needs and the constraints of the stakeholders interested in using the system.

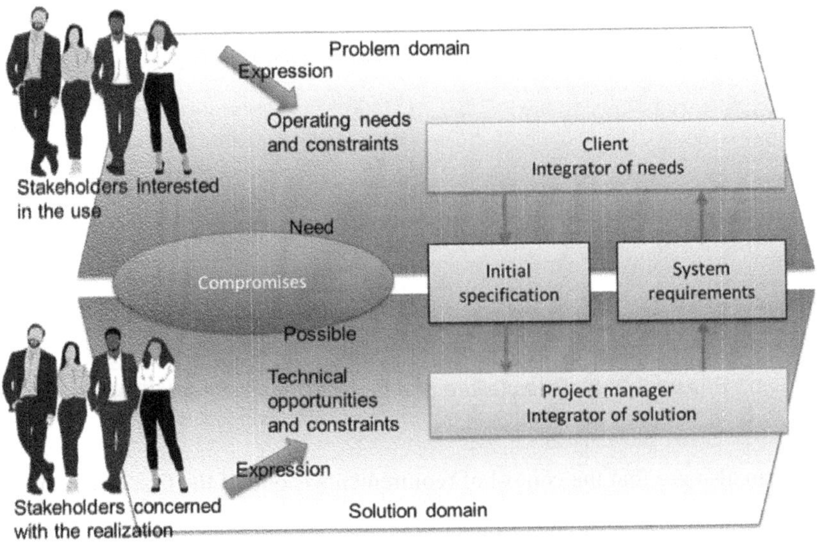

Figure 3.3. *Problem and solution domain*

The solution domain is covered by the project manager, who is the manager and integrator of the solution. The project manager takes into account the specifications of the project owner and captures the needs of the stakeholders involved in the realization of the system.

These stakeholders express the technical opportunities and constraints of the solution. The project manager formalizes the response to the needs of the initial specification by taking into account the technical opportunities and constraints in a system specification containing the system requirements.

There is a trade-off between the system's needs and requirements. The result must be validated by the project owner. The system requirements represent the contract between the owner and the developer and **define the problem to be solved**.

Figure 3.4 presents all the requirements engineering activities that are linked to the definition of the requirements starting from the needs to the specified requirements through the subsystems which will have been selected for the solution.

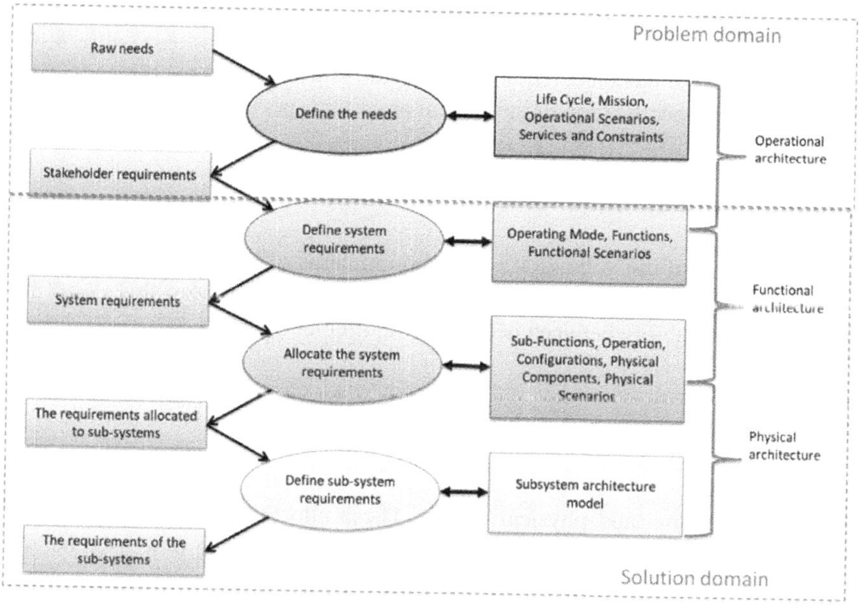

Figure 3.4. *Requirements engineering activities*

These activities are divided according to the problem domain or the solution domain. Each activity is defined by the input and output requirements (Hull et al. 2011). Next to each activity, we find the system architecture concepts that are defined in parallel. We also made the connection with the three architectural visions.

3.4. Formalizing stakeholder needs and system requirements

A requirements repository is a structured collection of requirements that can be managed in configuration. For the development of a system, the project manages two requirement repositories, namely (AFIS 2012):

– the initial requirements repository which expresses the needs of the stakeholders interested in using the system;

– the system requirements repository which refers to the initial requirements by adding the constraints of the selected system concept. This repository must be validated by the client.

The collection of stakeholder requirements can be put into a document called a technical specification of need or specifications.

The collection of technical requirements of the system can be put into a document called a technical specification of the system or system specification.

3.5. Validate system requirements

The objective of the validation of the requirements of the system is to see if the problem posed corresponds to the needs. The prerequisite for this is to have bidirectional traceability between needs and requirements.

3.6. Allocate system requirements to subsystems

Allocated requirements are system requirements allocated to the subsystems during the functional and physical design. These allocated requirements form the initial requirements of the subsystems. The systems engineering process is iterative. Indeed, the initial requirements are the requirements of the stakeholders for the subsystems.

3.7. Validate the allocated requirements of the subsystems

The allocated requirements must be validated to demonstrate that they address the problem posed by the system requirements.

3.8. Choose the solution: define the requirements of the subsystems

The specified requirements of the system and its components are formalized at the end of the design when the system definition is validated. These specified requirements form the reference for the entire life cycle of the system and its components (produce, integrate, verify, implement, maintain in operational condition and withdraw from service according to the EIA-632:1999 standard (ANSI 1999)).

3.9. Managing requirements

3.9.1. *Why manage requirements?*

Requirements management is part of the activities related to project management. It is essential to manage your requirements repository well, as it will inevitably evolve during the course of the project. There are several reasons for this. Here are three that we think are important. The first reason is the highly iterative nature of systems engineering activities. It is not realistic to wait for full validation of an activity before starting another one. For example, the design of the system can begin when not all requirements are validated.

The second is the maturity of the project owner's request. It is common that the client is in the reflection phase and that their needs are not all at the same level of maturity or all defined. The problem of maturity of the client's needs is common in projects involving technological breakthroughs.

The third reason is the maturity of the project manager's project. Indeed, at the beginning of the project, a set of technological paths is open and seems to meet the requirements. As the project progresses in the design, new constraints appear and avenues are closed.

3.9.2. *What are the risks associated with poor requirements management?*

If the requirements repository was not managed or was poorly managed, it could be a major source of error for the project. Here are some of the most classic ones. The project could have multiple teams or vendors working on different repository versions. This management error is a major source of delay for the project as unnecessary or false activities will be carried out. Moreover, this error will also generate additional costs for the project, as all or part of the activities that were carried out with the wrong version of the repository will have to be repeated.

The project did not or did not adequately realize the traceability between the requirements and the design elements. This project would not be robust to change as it would be unable to do an impact analysis related to a request for a change in a requirement. Moreover, without traceability, the validation activities of the requirements repository become impossible.

The project does not have a life cycle on these requirements. It is impossible for the project to distinguish the different states of maturity of its requirements. The project cannot properly manage changes in its repository.

3.9.3. *Managing requirement attributes*

In order to manage a requirements repository, it is necessary to associate a set of attributes with each requirement. We will present the minimum list of attributes for good requirements management. This list is not exhaustive, but we can recommend that the project reduce the number of attributes to the bare minimum to avoid wasting time and losing readability.

A requirement must have at least:

– An **identifier**: this unique number, valid for the entire life of the project, uniquely identifies the requirement. This attribute will also be used for traceability.

– A **description**: this text is the body of the requirement. The requirements are managed in a black box in the form of free text. We will see later that we can introduce a methodology to make the requirements semi-formal, thanks to a common structure.

– A **version**: this number should identify major and minor versions of the requirement. This attribute is essential for requirements configuration management.

– A **life cycle**: the life cycle value gives the state the requirement is in. This attribute is essential for change management.

– **Upstream and downstream traceability between requirements**: these traceability links allow the requirement to be linked to other requirements. For upstream traceability, we will speak about the coverage link. For downstream traceability, of refinement or declination links.

This proposed list is the minimum for good requirements management. To facilitate the management of the project, we can add the following attributes:

– A **name**: adding a name to the requirement makes it possible to summarize the requirement. This makes it easier to read the document or database.

– **Validation**: this attribute makes it possible to select the method that will be used to validate the requirement.

– A **verification**: this attribute makes it possible to select the method that will be used to verify the requirement on the embedded system.

– An **allocation**: this attribute allows the requirement to be allocated to a system architecture element.

This additional list is not exhaustive. The reader is invited to consult AFIS (2012) and Stéphane and Badreau (2014) for more information.

3.9.4. *Managing the configuration of requirements*

The configuration management of a requirement repository consists of being able to identify, in an unambiguous way, a set of requirements in a certain version during the course of the project.

3.9.4.1. *Manage requirement versions*

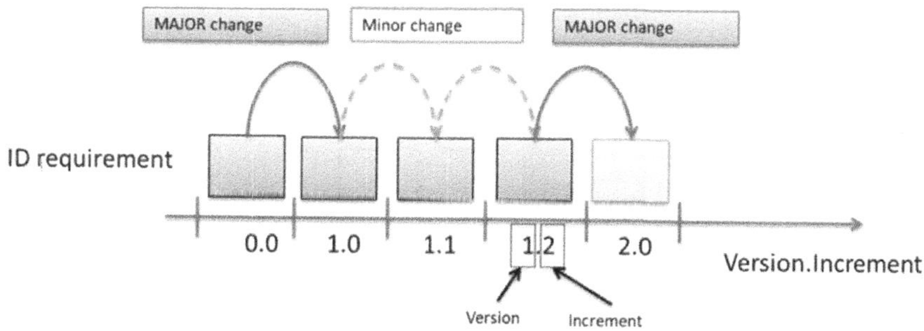

Figure 3.5. *Version of a requirement (Stéphane and Badreau 2014)*

In order to manage requirements, it is essential to be able to unambiguously identify the version in which it is located (see Figure 3.5). A requirement version can be composed of the version itself and an increment. The version represents the major evolutions of the requirement and the increment the minor evolutions.

3.9.4.2. *Defining a requirements configuration*

A requirements configuration is defined by two dimensions, namely a "product" dimension to have the set of unitary requirements that characterize the system and a "version" dimension to unambiguously identify the different versions of the requirements (Stéphane and Badreau 2014).

For example, in Figure 3.6 we have a requirement 1 configuration with requirement 1 in version 3, requirements 2, 4 and 5 in version 1 and requirement 3 in version 2.

A requirement configuration is a snapshot of the repository and should be characterized by the following elements:

– it must have a defined field of application: for example, specification of the system or subsystem;

– it should be as consistent as possible;

– it must not be modifiable;

– it must be identified.

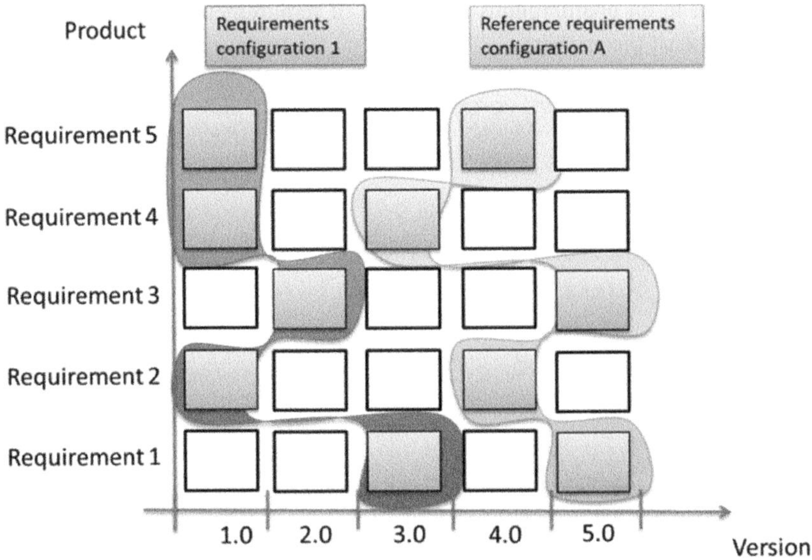

Figure 3.6. *Requirements configuration (Stéphane and Badreau 2014)*

In a project, not all configurations are equally important. We have the baseline or reference configuration. In general, a requirements reference configuration corresponds to a project review or a prototype (see Figure 3.7).

We define a reference configuration by a set of documents that is versioned. A requirements reference configuration is a set of specification document versions containing requirements frozen in a given version.

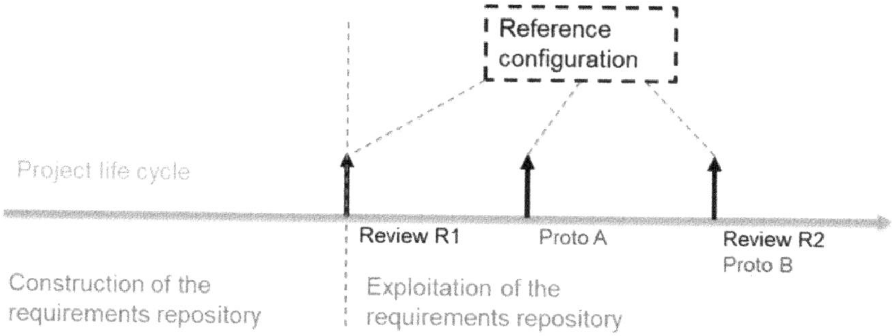

Figure 3.7. *Reference configuration*

For example, in Figure 3.6 we have a reference configuration A with requirements 1 and 3 in version 5, requirement 4 in version 3 and requirements 2 and 5 in version 4.

In reality, the requirements repository is often built in an incremental way where the first reference configurations are often partial.

3.9.5. *Managing changes*

As stated in the introduction to requirements management, changes during a project are almost inevitable. For this, the requirements management process must allow for change management. In order to manage changes, we need to be able to perform impact analyses and manage **a life cycle for each requirement**.

3.9.5.1. *Managing traceability*

There are several types of traceability in systems engineering. We have traceability between requirements, which can be on several or the same systemic level. We also have traceability between requirements and design elements such as functions or system components.

3.9.5.2. *Managing the life cycle*

A requirement from its creation to its release or abandonment goes through several states. These states form the requirement life cycle and are managed by the requirements management process (see Figure 3.8).

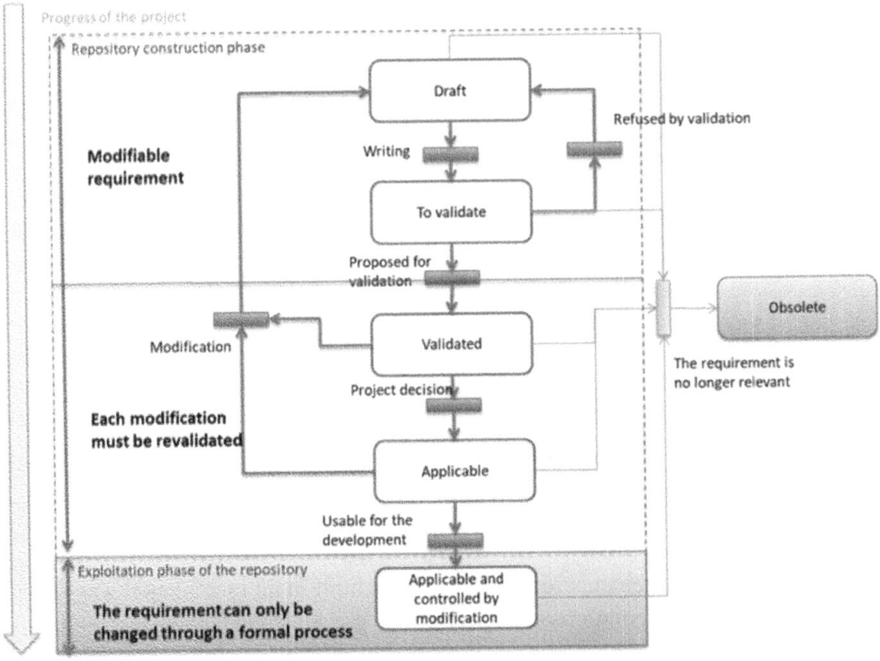

Figure 3.8. *Requirement life cycle*

In Figure 3.8, we have a requirement life cycle broken down into three phases which correspond to the course of the project. We have the construction phase of the repository where the requirements can be modified. Then, we have the repository construction phase where any changes must be revalidated. And finally, we have the operation phase of the repository where the requirement can only be modified through a formalized process.

3.10. Conclusion on requirements engineering

Requirements engineering has its own activities on all the processes of systems engineering. During the course of these processes, the requirements decline and change.

Indeed, during the analysis, the needs are formalized into stakeholder requirements. During the specification, the stakeholders' requirements are translated into system requirements. During functional and physical design, the system requirements are allocated to the constituents to become the initial requirements of the subsystems. After the evaluation and comparison of the architectures, the initial requirements of the subsystems become the specified requirements to enable the realization of the solution elements.

In order to manage a requirements repository in configuration and to manage evolution requests, the project must set up bidirectional traceability between needs and requirements and manage attributes on the requirements.

For example, for traceability and configuration management, requirements must have a unique identifier.

For configuration management, the requirements must also be versioned. And for change management, the project must implement a life cycle on the requirements.

PART 3

Definition of Requirements

4

Unifying Thread Example

4.1. Objective of the unifying thread example

To illustrate the systems engineering approach, we will work through a unifying thread example for the design of a water pump with a wind generator. This example is fictitious, has an educational purpose and is not representative of the state of the art in the field of pumps and wind turbines.

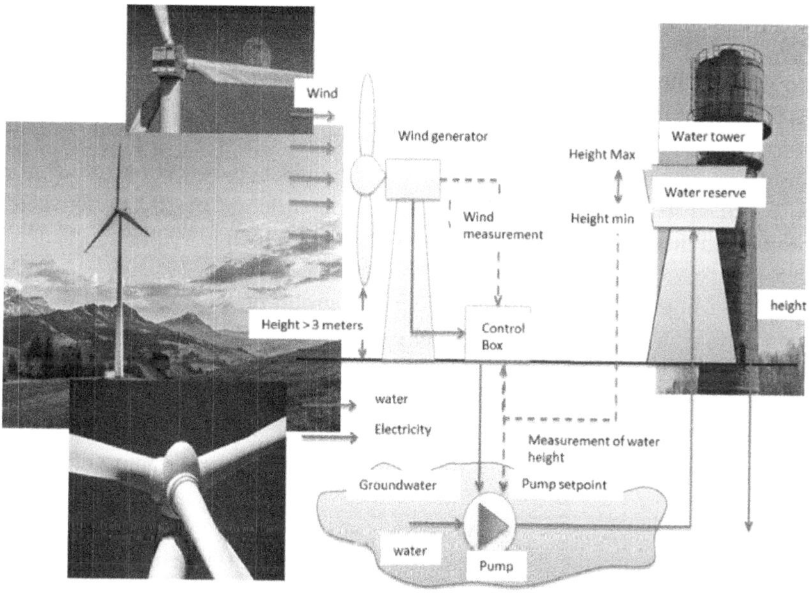

Figure 4.1. *Water pump with a wind generator*

For a color version of all the figures in this chapter, see www.iste.co.uk/hygounenc/system.zip

4.2. Presentation of the raw specifications

A municipality wishes to equip its water tower with a new water pump that uses renewable energy. Europe offers aid for the development of this type of technology, up to 50% of the recurring costs.

The geographical position of the municipality has windy weather that allows the use of wind technologies. In fact, the weather imposes winds between 0 and Vmax = 8 m/s.

The municipality therefore launched a call for tenders to build a water pump using wind energy. The residents want, when the water tower is full and the weather allows it, the electrical energy produced by the wind turbine to be sold to the electrical network in order to reduce their water bills.

The electricity grid agrees to buy back the kwh at €0.5, but imposes that the electrical energy should be in 220 Vac and 50 Hz. The municipality requires the supplier of the solution to ensure continuity of service for filling the water tower even when there is no wind. The solution provider will be able to use the electrical grid as an alternative energy resource.

The water table is abundant with a low calcareous water, with a hardness HT < 10°F (HT = hydrotimetric title, °F = French degree).

The water tower must always have a water height between Hmin = 3.5 m and Hmax = 5 m in its tank. The water tower requires a height of 100 m between the water table and the top of the reservoir, the volume of which is 300 m^3.

After the filing of the construction permit, a group of local residents expressed their concerns and wanted the noise level of the wind turbine not to exceed 45 dB when the wind was at 8 m/s.

From a designer's point of view, their wind turbine supplier offers them a wind generator capable of producing electricity with wind between Vmin = 3 m/s and Vmax = 8 m/s.

Figure 4.2 presents the boundaries of the system in green and its external environment.

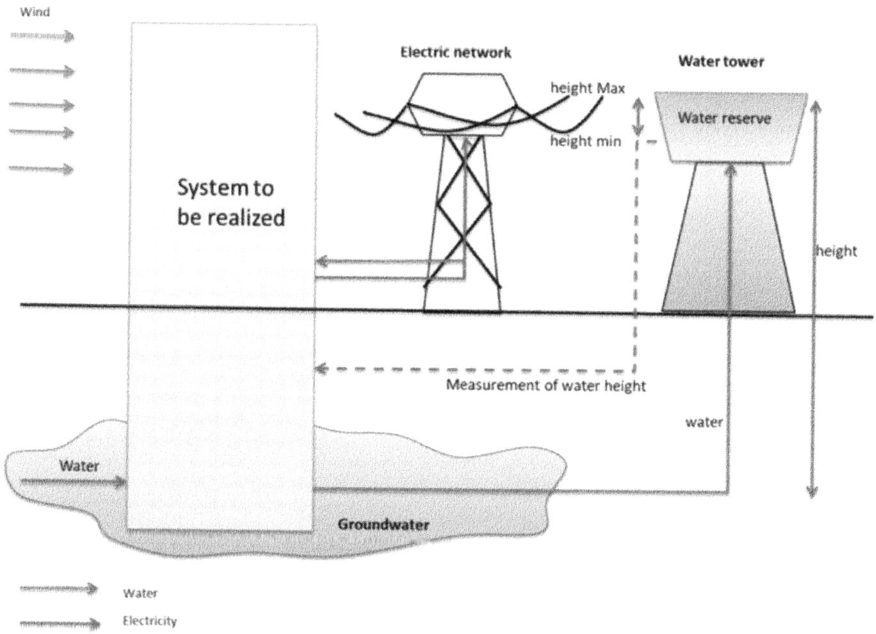

Figure 4.2. *System environment*

5

Needs Analysis

5.1. Objectives of needs analysis

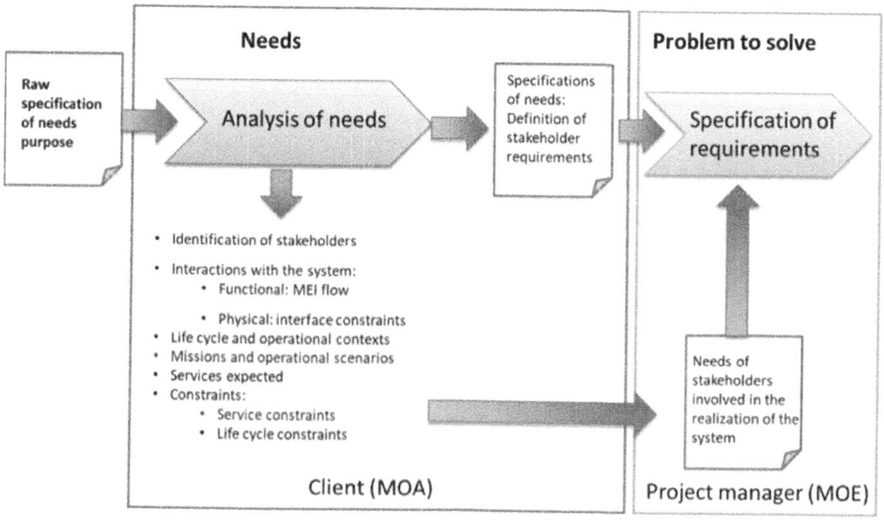

Figure 5.1. *Needs analysis*

The objective of needs analysis is to identify the needs of the problem to be solved and to formalize them in the specification of needs in the form of stakeholder requirements (see Figure 5.1). To conduct this analysis, we will:

– identify stakeholders;

– model the interactions between the external environment and the system;

For a color version of all the figures in this chapter, see www.iste.co.uk/hygounenc/system.zip.

– build the life cycle and identify operational contexts;

– identify the missions of the system that can be put in the form of use cases;

– write the operational scenarios of the system;

– identify the expected services and the constraints imposed by the environment;

– repeat and formalize the needs in stakeholder requirements.

5.2. Stakeholder identification

5.2.1. *Definition of stakeholder*

The stakeholders are people or organizational entities with expectations or constraints regarding the system to be realized. We have two types of stakeholders: those who are directly interested in the use of the system and those who are concerned with the realization of the system (Figure 5.2).

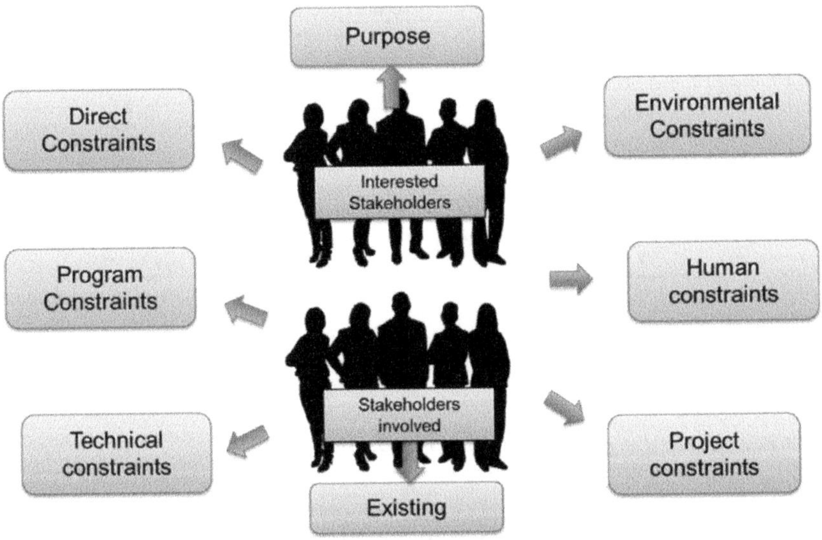

Figure 5.2. *Stakeholders and their needs*

Stakeholders interested in the operation and use of the system express:

– the use, i.e. the purpose, through the missions and the expected performance;

– the direct constraints of the system with the definition of the interactions with external systems;

– environmental constraints;

– human constraints.

Stakeholders involved in the realization of the system express:

– opportunities to reuse existing and market products;

– the business opportunities of the system or other purposes according to the company's aims;

– program constraints on the life cycle of the system (deployment, operation, logistical support, maintenance, retirement);

– human constraints on skills, recruitment and training;

– development project constraints on design, integration, production;

– technical constraints on the state of the art, trades, subcontractors and standards.

All of these stakeholders and external systems together form the system's environment. In this environment, we have:

– stakeholders and external systems that have a direct impact on the design of the system because they use it or interact with it directly;

– stakeholders who have an indirect impact on the company's policy, regulations, competition, etc.

5.2.2. *PESTEL analysis*

To identify the stakeholders who are affected by the realization of the system or who have an indirect impact on the design of the system, we can use the PESTEL analysis, for political, economic, societal, technological, ecological and legal (see the table in Figure 5.3).

This analysis is used as a guided brainstorming where we can use the six PESTEL themes as an aid to identify stakeholders.

For example: Are there political stakeholders or those who have political interests in the system to be realized?

This analysis can also be used as a technique for classifying stakeholders in the environment. This gives an idea of the completeness of the stakeholder identification.

Types of needs and constraints of a system		External systems
P	Policies	Political system
E	Economic	Economic system
S	Societal	Society
T	Technological	Technology Innovation System
E	Environmental	Earth System
L	Legal or regulatory	Legislative and regulatory system

Figure 5.3. *PESTEL analysis*

5.2.3. *Analysis by type of environment*

PESTEL analysis is not sufficient to identify the stakeholders who are part of the direct or indirect environment of the system. Figure 5.4 breaks down the system environment into five sub-environments. First, we have the "to be realized" system which interacts with the "to be done" system which we will call the project environment. The project is in interaction with the client's environment and the normative environment. These three environments represent the indirect environment of the system. The client's environment interacts with the environment of the system's users. The normative environment interacts with the environment in which the system is used. The environment of the users and use interact with the system to be made and are part of its direct environment.

These environments are sources of needs or they have impacts on the design of the system to be achieved.

The project environment includes the organization of the project with suppliers and the company. This environment is a source of opportunities such as the use of equipment developed in R&D or the establishment of a partnership. This environment is also a source of constraints on costs, performance and deadlines.

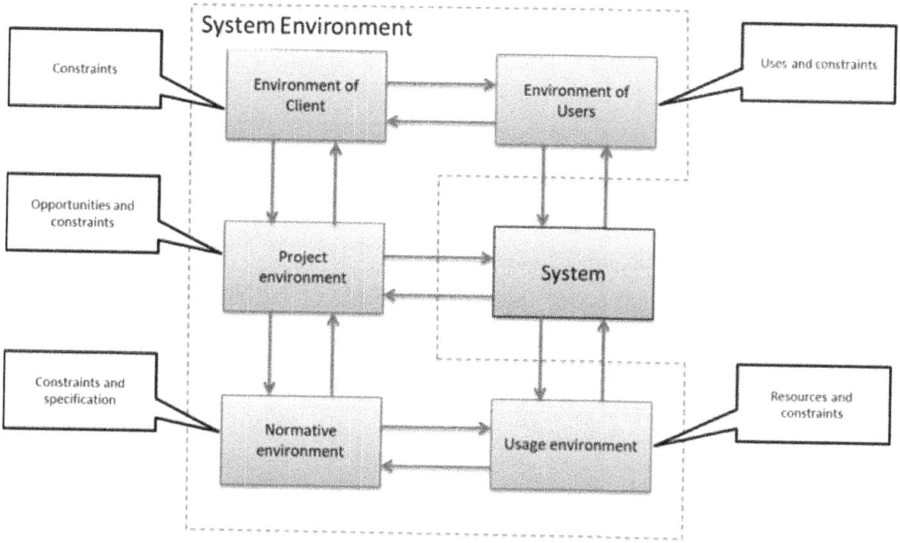

Figure 5.4. *The system environment*

The client's environment includes the client's organization and the competition. This environment is a source of constraints such as the price of the system.

The normative environment or other stakeholders include standard bodies such as European agencies. This environment is a source of constraints such as the restriction in the use of certain materials. This environment can also be a direct source of requirements for the system to be made.

The user environment consists of the external stakeholders and systems that have use of the system. This environment can also impose constraints such as the human factor or performance.

The environment of use is made up of all the external systems that interact directly with the system without using it. This environment is a source of constraints but can also serve as a resource for the system. It can be divided into two sub-environments, namely the natural environment and the induced environment (see Figure 5.5).

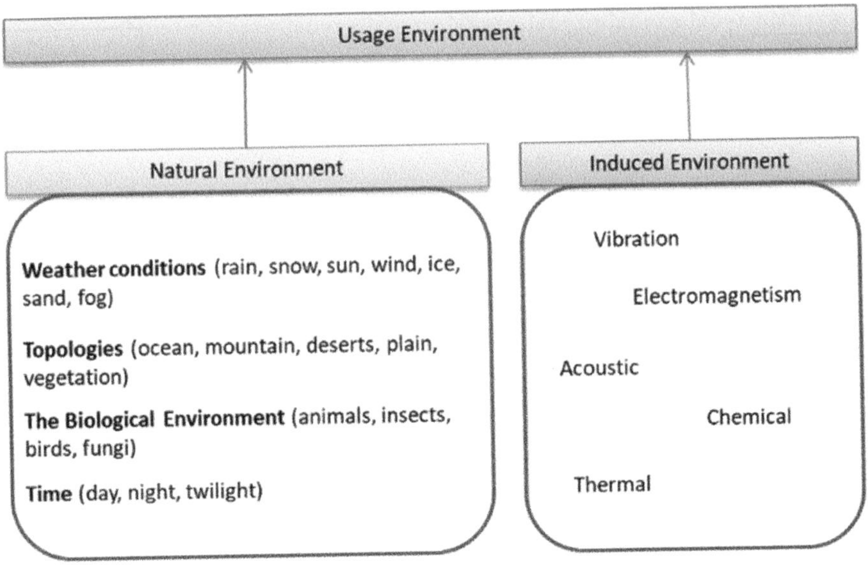

Figure 5.5. *Environment of use*

5.2.4. *Application to the unifying thread example: stakeholder identification*

The analysis of the "raw" specifications on the renewable energy water pump allows us to identify a number of stakeholders and external systems. From our raw specifications, we will identify external stakeholders or systems. For example, in this excerpt, we have highlighted the stakeholders we identify.

> [...
>
> *The municipality therefore launched a call for tenders to build a water pump using wind energy. The residents want, if the water tower is full and the weather allows it, the electrical energy produced by the wind turbine to be sold to the electrical network.*
>
> ...]

The following table summarizes the analysis by proposing a PESTEL classification if relevant and the position in one of the five environments.

Stakeholder	Note	PESTEL	Environment
Municipality	It is the customer who buys the solution, but it also has a political interest	Policy	Customer
Water tower	It is the user of the system	Technological	User
European Union		Legal	Normative
Weather	It is a system resource		Use
Administered		Societal	Customer
Electrical network	It is both a resource and a user of the system	Technological	User
Solution provider	We also find the designer of the solution in the specifications	Economic	Project
Water table	It is a system input that contributes to the value of the system		Use
Collective of residents		Societal	Customer
Wind turbine supplier		Economic	Project

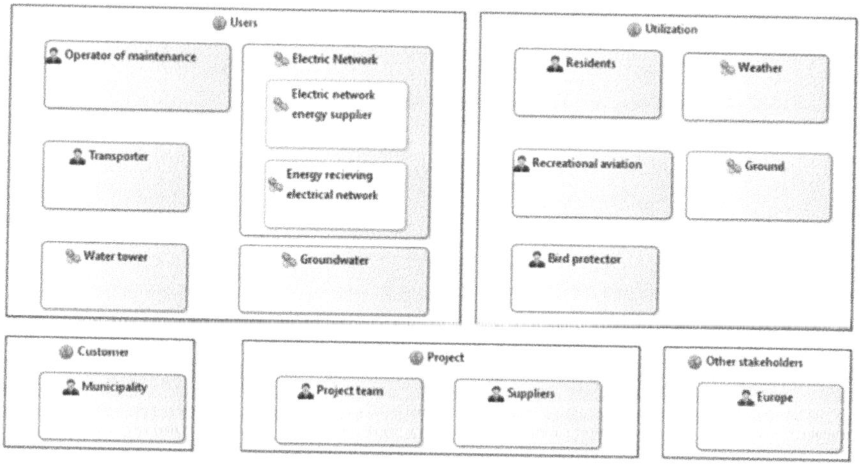

Figure 5.6. *Environment of the unifying thread example*

By completing the analysis of the specifications through brainstorming, we can finalize the identification of the environment. Figure 5.6 presents the environment of the hierarchical system. We can note that the electrical network is presented in the form of its roles vis-à-vis the system to be achieved; indeed, we have the electrical network which has a role of energy supplier and a role of energy purchaser. This formatting is called an environment map and is one of the first deliverables in systems architecture. We see that it is a tree model.

5.3. Identification of external interactions

5.3.1. *Objectives of identifying external interactions*

From the direct environment, i.e. the environment of users and use, we will identify the interactions of energy and information with the system seen as a black box (see Figure 5.7). We will also talk about MEI flow exchange.

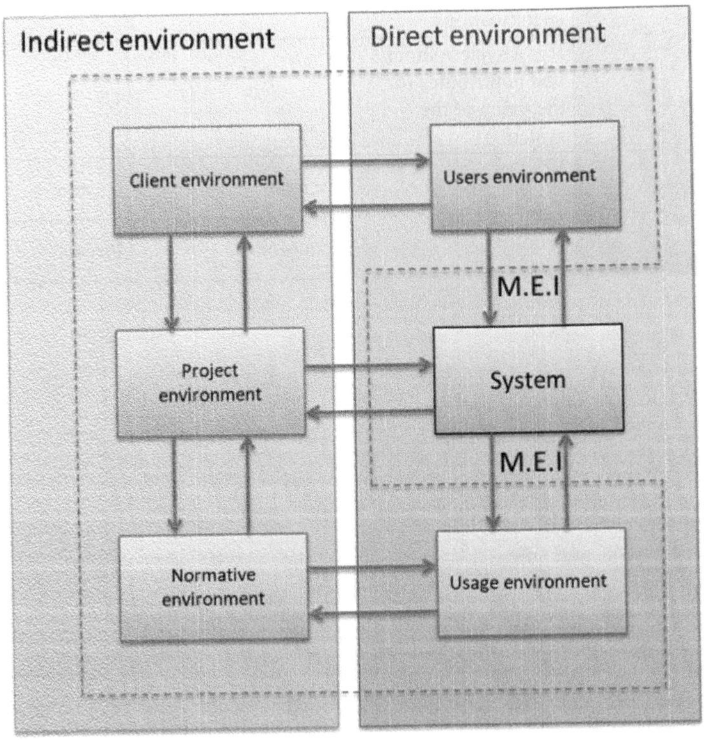

Figure 5.7. *Environment interacting with the system*

This key activity has several objectives:

– identify the perimeter of the direct environment: significant environment;

– identify the perimeter of the system to be created, i.e. its boundaries or limits;

– identify the external interfaces: functional and physical;

– identify the departure and arrival of future functional chains: transformations of incoming flows into outgoing flows.

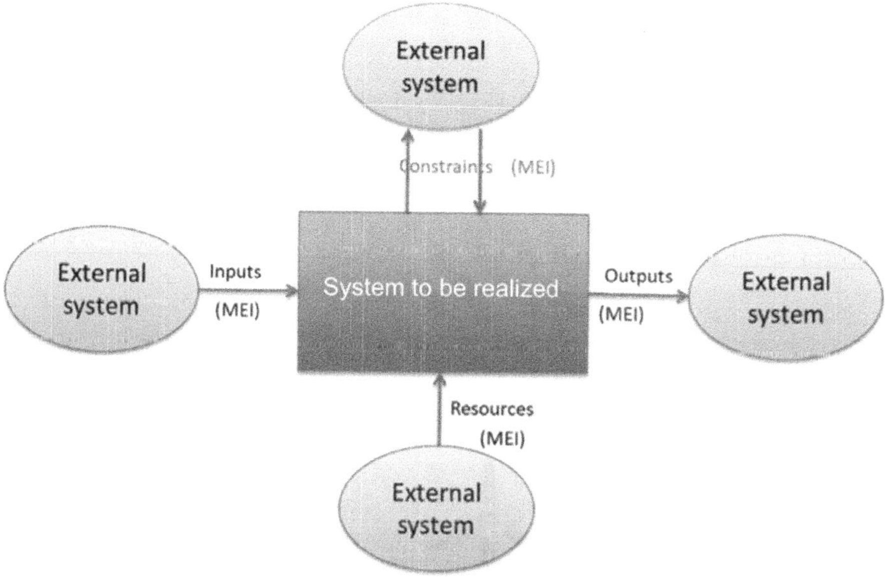

Figure 5.8. *Block diagram of external flows*

The interactions between the system and the environment can be represented as a block diagram (see Figure 5.8). To construct this diagram:

– we will identify the purpose of the system, i.e. the MEI (matter, energy and information) output flows that will be consumed by the environment;

– then, we will go back to the MEI input flows that allow the system to transform these output "goal" flows. This is the system's operating chain;

– to achieve this value chain, the system relies on input resources. These resources are technical flows that can be the choice of the project manager;

– we can complete this by identifying the input and output flows that allow the environment to drive the system;

– finally, we can identify the MEI flows from the environment, which do not contribute to the value but will constrain the system. Depending on the technical choices made by the project manager, MEI flows may leave the system and constrain the environment.

5.3.2. *Application to the unifying thread example: identification of external interactions*

We can build a block diagram from the user and use environment. For each stakeholder or external system, the objective is to identify the MEI flows (matter, energy and information) exchanged with our "renewable energy water pump" system (see Figure 5.9).

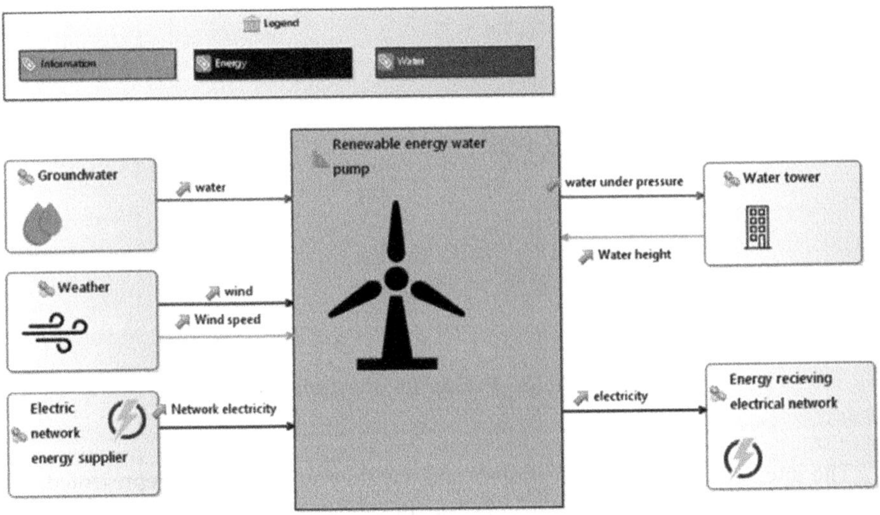

Figure 5.9. *Environmental flow map*

Following the method in section 5.3.1, we identified the output streams that are the goals of the system. In this example, the system must supply the environment with pressurized water and electricity (see Figure 5.10). Note that this map is static. It does not model the fact that the two output streams are not produced at the same time.

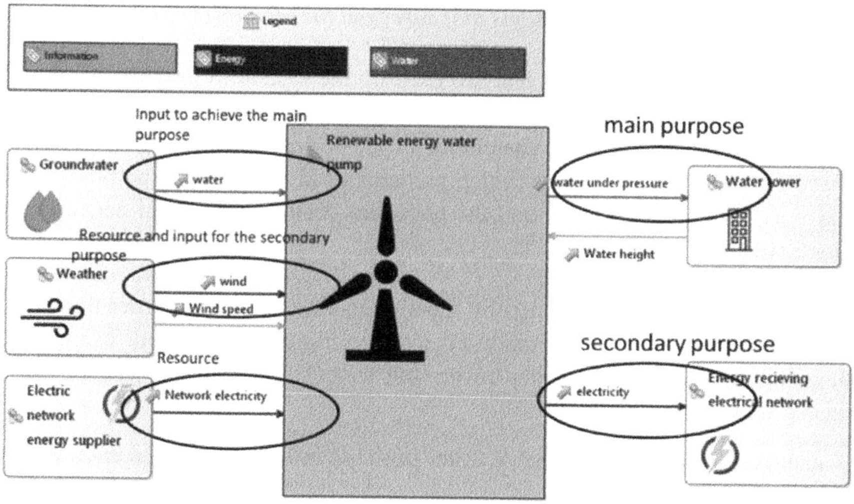

Figure 5.10. *Definition of the inputs and outputs of the system*

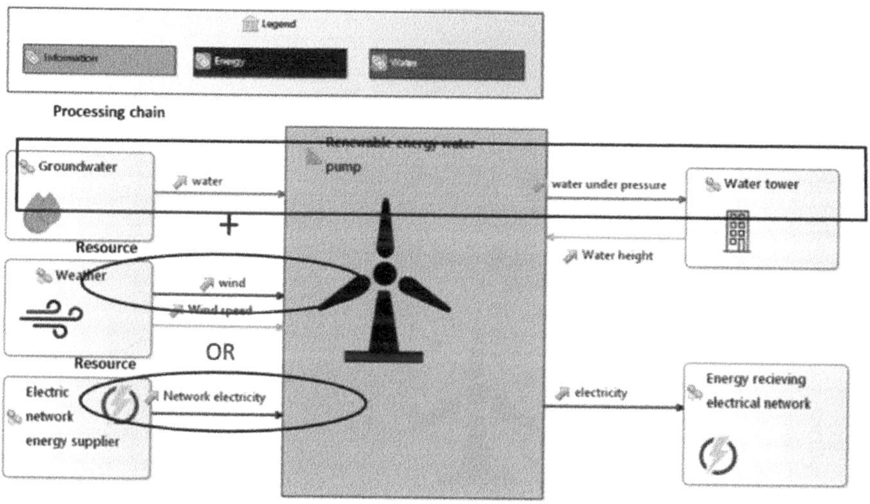

Figure 5.11. *Chain of operation*

What are the chains of operations that carry out the missions?

Then, we identify the inputs that are used by the system to provide the outputs that will be consumed by the environment. Among these input flows, we have the **resources** which are most often the choice of the project manager. In our example, we have groundwater as an input which participates in the system's purpose. Then, we have as a resource the wind and the electricity from the electrical network (see Figure 5.10).

The inputs that are transformed by the system into outputs, in order to achieve the system's goal, form the operation chains (see Figure 5.11). In our example, to achieve the mission "fill the water tower with water", the system must (see Figure 5.11):

– chain of operation 1: pressurize water from groundwater and wind energy;

– chain of operation 2 (alternative): pressurize water from the water table and energy from the electrical grid.

For the second mission, the system takes wind energy as input and transforms it into electricity for the power grid as output.

How is the system controlled?

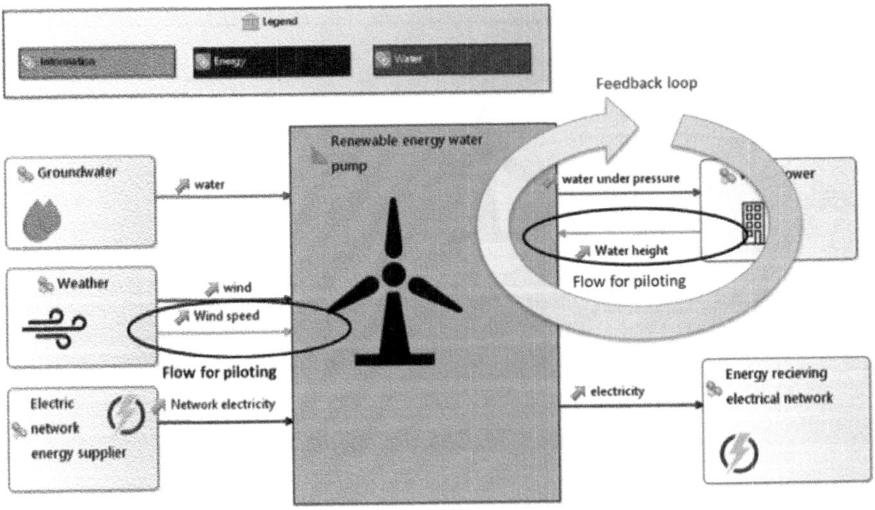

Figure 5.12. *Steering chain*

Finally, we have also identified in this map the flows that will be used to drive the system. We have two input streams that will be used to regulate and change the operating modes of the system which are the water height in the tank and the wind speed measurement (see Figure 5.12).

We can identify a feedback loop on our system. We have a loop between the water tower and the system where the head of water is the cause that will cause a flow of water, which is the effect that will modify the cause which is the head of water.

5.4. Collection of needs

5.4.1. *Why capture needs?*

In the definition of systems engineering, it was stated that this approach serves to propose a solution that meets identified needs. It is not sufficient to consider only the needs that are in the specifications of the client. Indeed, a customer never writes down all their needs. To explain this, we can use the Kano diagram (see Figure 5.13).

On the x-axis, we have the degree of implementation of the functional or constructional feature on the system. On the right it is present and on the left it is absent. On the y-axis, we have customer satisfaction. At the top, they are satisfied and at the bottom they are dissatisfied. The **Kano diagram** is based on three curves.

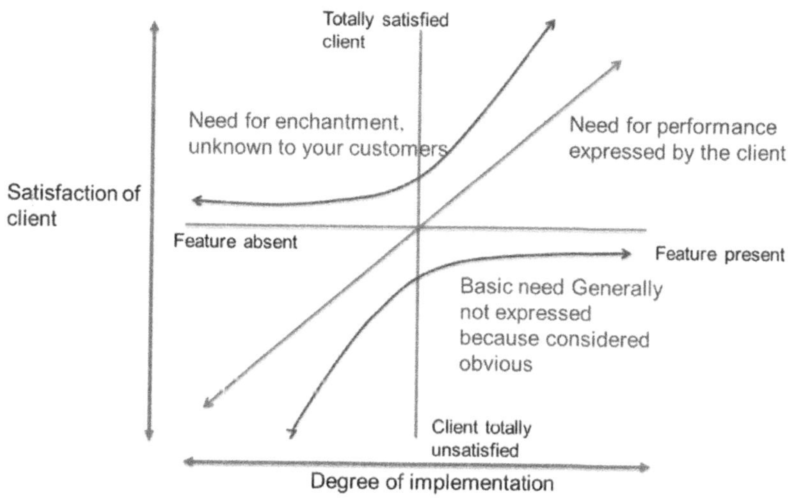

Figure 5.13. *Kano diagram*

The **red curve** represents the basic requirement. We can note that if this need is not covered by the system, the customer's dissatisfaction is immediate. The basic need is often not written in the specifications by the client because it is obvious to them. However, what is obvious to the client is not necessarily obvious to the contractor, hence the possible conflicts during acceptance of the integrated system.

The **blue curve** represents the performance requirement. Generally, this is the need that is expressed in the specifications.

The **green curve** represents the need for enchantment. This need may be written into the specifications, but it may be proposed by the project manager as it can be an effective way to differentiate from the competition.

The Kano diagram can be a method for classifying and prioritizing requirements.

5.4.2. *Methods for collecting needs*

The collection of needs can be done through four methods, namely:

– analysis of input documents such as the project owner's specifications;

– market research;

– brainstorming;

– stakeholder interviews.

From the list of stakeholders, we can associate one of these four methods for collecting needs. This association can be put in the form of a needs assessment plan. For each stakeholder, identify:

– **required capabilities**: i.e. what the stakeholder wants or requires;

– **operational contexts**: i.e. under what conditions the capability is required;

– **expected performance**: i.e. what are the quantifiable criteria that allow us to know if the capacity is met.

Requirements can be formalized using the canonical forms presented in the table in Figure 5.14. These requirement templates allow us to capture and format for each stakeholder the expected **services** and the **constraints** imposed on the system.

Need template	Template
Service expected by a stakeholder	The <Stakeholder> **wants** <something> with a <target performance level> in a <given context>
Constraint imposed by a stakeholder	The <Stakeholder> **imposes** <something> with a <target performance leve>l in a <given context>
Demanded capacity by an external system	The <external system> **must be (not be) capable to do** <something> with a <target performance level> in a <given context>

Figure 5.14. *Need template*

5.4.3. *Application to the unifying thread example: formalization of needs*

The application to our unifying thread example will be done in two steps. First, we analyze the raw specifications to extract for each stakeholder or external system the following elements:

– services, capacity requested or constraints imposed;

– the level of performance or the associated quantitative satisfaction criteria;

– the operational context during which the capability is requested or the conditions of activation.

Second, we format these needs in a four-column table with:

Applicant	Capacity	Criterion	Context

Applying this approach to a part of the "renewable energy water pump" specification, we obtain the following result:

A municipality wishes to equip its water tower with a new water pump that uses renewable energy. Europe offers aid for the development of this type of technology.

The geographical position of the municipality has windy weather that allows the use of wind technologies. Indeed, the weather imposes winds between 0 and Vmax m/s.

The municipality therefore launched a call for tenders to build a water pump using wind energy. The residents want, if the water tower is full and the weather allows it, the electrical energy produced by the wind turbine to be sold to the electrical network.

The electrical network accepts, but imposes that the electrical energy is in 220 Vac and 50 Hz. The municipality requires the supplier of the solution to ensure continuity of service for filling the water tower even if there is no wind. The solution provider will be able to use the electrical grid as an alternative energy resource.

...

***The water tower must always have a water height in its tank between Hmin and Hmax. The water tower has a height of 100 m** and a volume of **300 m**^{3}.*

We format all of these needs in a table and use the templates for reformulation. When an element is missing, we put a question mark. A reminder that one of the objectives of systems engineering is to make it possible to work in a state of uncertainty, but knowing where we are, i.e. what remains to be done.

Applicant	Capacity	Criterion	Context
The municipality	Wants to equip itself for the supply of its water tower	A renewable energy pump	?
The weather	Imposes winds	Between 0 and Vmax m/s	?
Residents	Want the energy produced by the system to be sold back to the grid	?	If the water tower is full and the wind >Vmin m/s
Electrical network	Imposes that the energy produced	Either in 220 Vac and 50 Hz	?
The municipality	Imposes on the solution provider a continuity of service	Water tower filled with water	Whether it is windy or not
The solution provider	Must be able to use electrical energy from the grid as an alternative resource	?	When there is no wind
Water tower	Must always be filled	With a water height between Hmin and Hmax	?
Water tower	Imposes a height	From 100 m	?
Water tower	Imposes a volume	From 300 m^3	?

We can see that there are gaps and that not all needs seem to be at the same level. However, it is easy to distinguish between services and constraints. To increase the maturity of these needs, an operational analysis is required. This analysis begins by identifying the system's life cycle.

5.5. Identification of the life cycle

5.5.1. *Building the system life cycle*

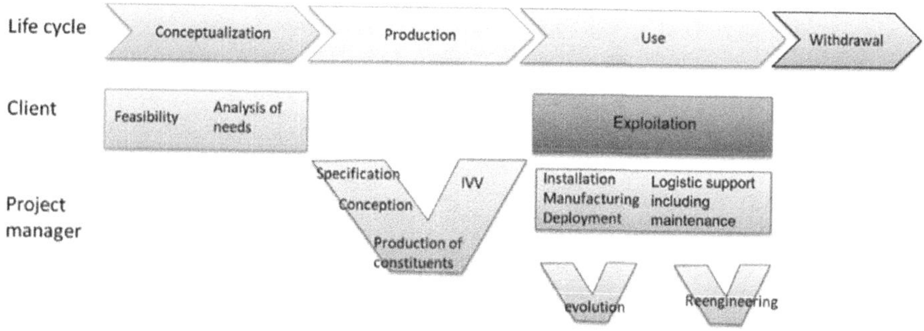

Figure 5.15. *System and project life cycle*

The life cycle of a system is composed of a set of operational contexts, each of which has an objective. This cycle (see Figure 5.15) covers the entire life of the system from the emergence of operational concepts to its retirement, including recycling (Meinadier 1998; AFIS 2014).

An operating context is a state of the environment unambiguously defined by a set of stakeholders and external systems (Krob 2017). Note that this definition of the environment also includes the system under study. This is why we speak of environment in the broad sense or external environment with the system.

In an operational context covered by the solution, the system itself is in a certain state, functional and constructional, to respond to services and constraints. The operational contexts in which the system renders a service are also called contexts of use of the system (de la Bretesche 2000).

For example, Figure 5.16 shows the life cycle of a civilian UAV with its different life situations and operational contexts. We can note that in the life situation "in operation", there are two concurrent state machines in order to avoid the combination of the contexts "flight phases" and "weather situation".

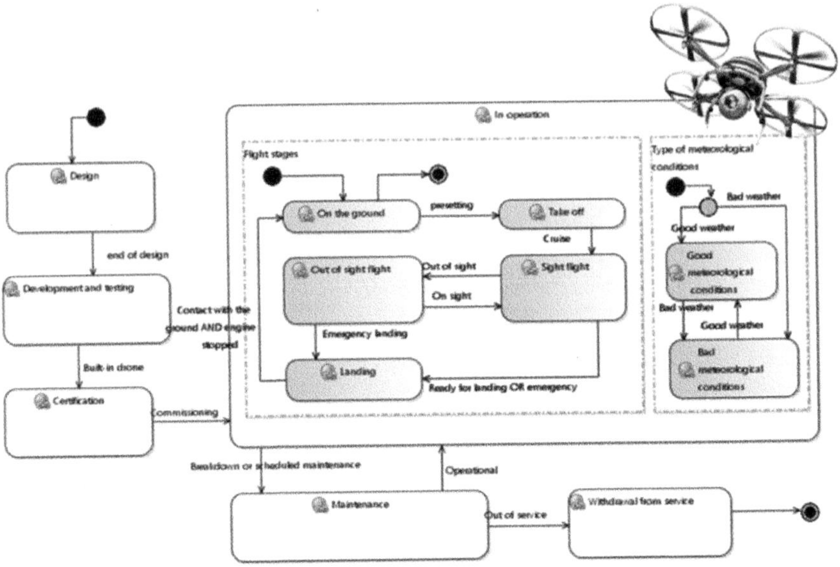

Figure 5.16. *Life cycle of a civil drone for aerial images*

The life cycle is built in three stages (see Figure 5.18).

– Identification of the operational contexts which are the main life situations of the system. These contexts are almost invariants for all systems. The life cycle in Figure 5.17 can be used as a starting point as a generic life cycle and can be adapted according to the context. It is indeed, in the main principles, the life cycle that we find in the standards ISO-15288 (ISO 2015), EIA-632 (ANSI 1999) and IEEE-1220 (IEEE 2017).

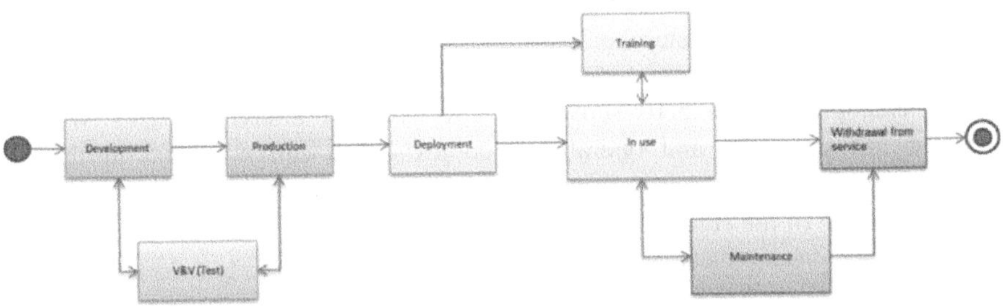

Figure 5.17. *Generic life cycle*

– Identification of the operational contexts which are the main contexts of use of the system. These contexts are specific to the purpose of the system and can be identified by analyzing the collection of needs and the states of the environment. Indeed, the "template" of the requirements defines the context of application of the services and the constraints. The set of external system and stakeholder states unambiguously formalizes the operational contexts.

– Identification of events between operational contexts. An event is a condition that makes it possible to move from one operational context to another.

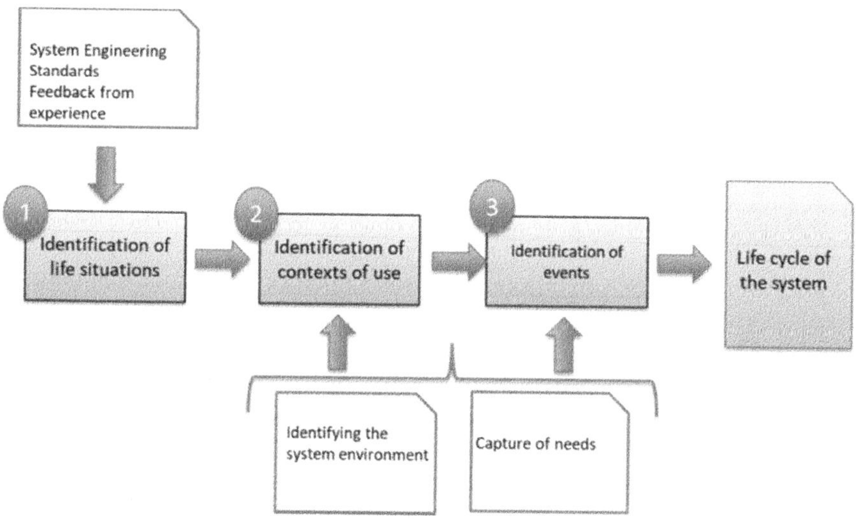

Figure 5.18. *Identification of the life cycle*

The life cycle of the system can be put in the form of a state machine. In this approach, we have chosen to use Moore's state machines, where the activation of services and functions is done in the state and not on the transition following the event (Villemejane 2013).

5.5.2. *Application to the unifying thread example: identifying the life cycle*

First-level operational contexts are life situations that are valid for all systems. Figure 5.19 represents the life cycle for our "renewable energy water pump" system. We can see that these life situations can be applied to any system. Only the sequence of some of these life situations can be specific to the system studied. By detailing the operational context "in service", we will identify all the contexts of use of the "renewable energy water pump" system. For this, we take as input the environment of the system as well as the collection of needs.

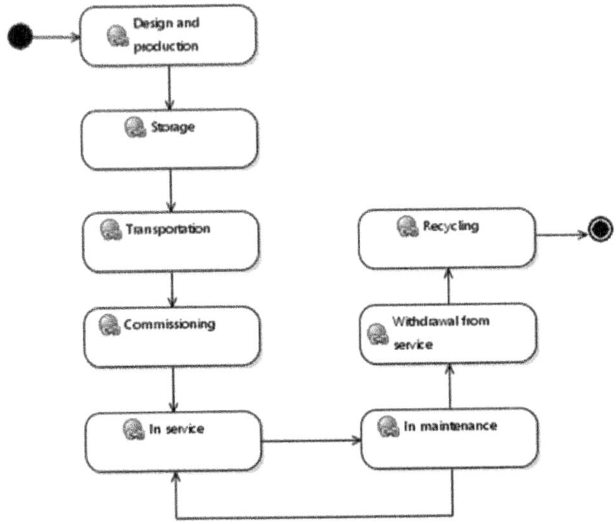

Figure 5.19. *System life cycle*

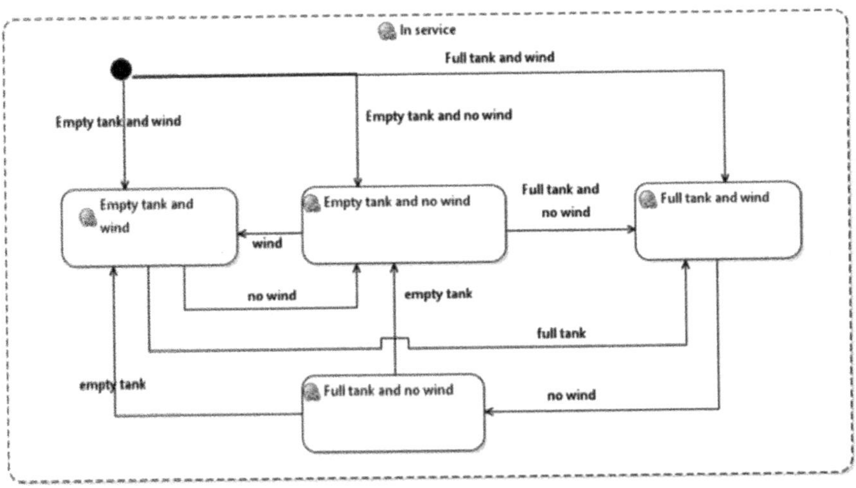

Figure 5.20. *Context of use*

In our thread, the contexts of use depend on the state of the water tower reservoir and the state of the wind in the weather. A method for identifying or verifying operational contexts is based on the combinatorics of environmental states.

To avoid too many operational contexts, related to the combinatorial explosion of external system states, we must:

– select only the largest external systems;

– associate with each of these external systems a list of restricted states, two or three at most at the risk of being caricatural (see Table 5.1).

This methodology can be completed and cross-referenced with needs analysis.

	Tank		Weather	
	Full	Vacuum	Wind	No wind
Context 1		X	X	
Context 2		X		X
Context 3	X		X	
Context 4	X			X

Table 5.1. *Method for identifying operational contexts*

5.6. Identification of the system's missions

5.6.1. *Definition of the system's missions*

The system to be achieved responds to an aim which is the main goal or the purpose of the system. This aim can be broken down into elementary goals. Missions are those goals that the system must achieve to meet the purpose (Meinadier 1998). These missions represent the possible uses of the system by the environment (Krob 2017) and are described and broken down into operational scenarios that encompass or link the services offered by the system (Faisandier 2012) (see Figure 5.21).

A mission transforms the state of the environment and can therefore change the operational context when it ends (Figure 5.22).

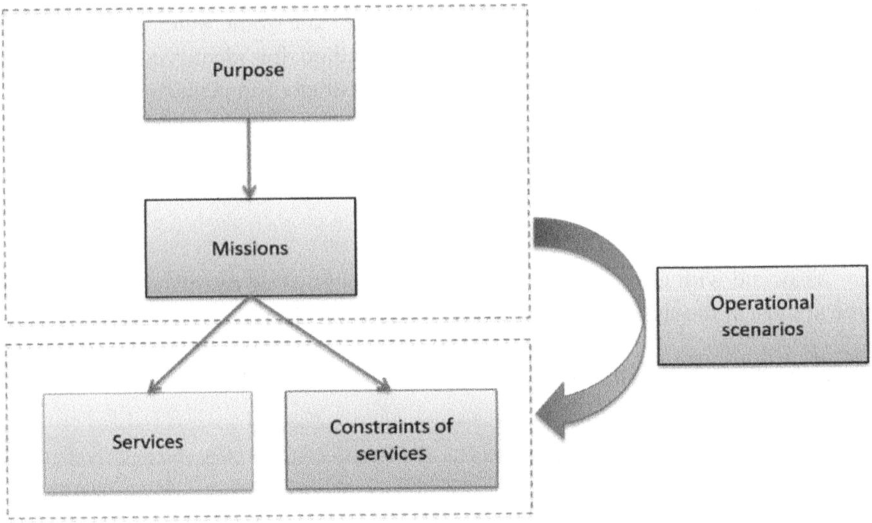

Figure 5.21. *The missions of the system*

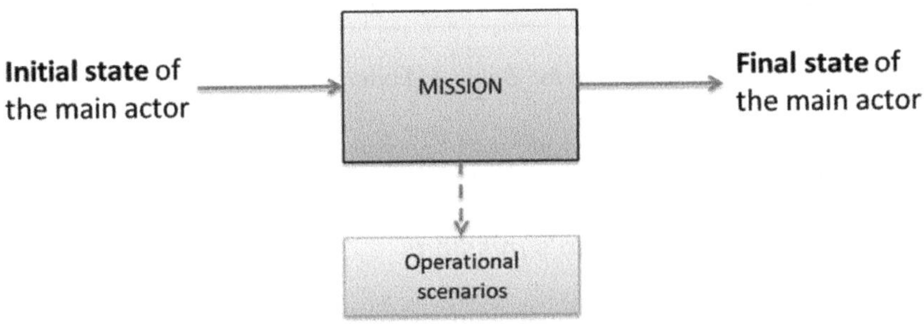

Figure 5.22. *The mission transforms the state of the environment*

To identify and characterize the missions, we must list the different uses that the environment will make of the system and put them into their operational context. For example, Figure 5.23 represents a Socoa semaphore poster on which we find in an explicit way all its missions. We understand that the French Navy is going to use the Semaphore to disseminate information, to monitor and to keep watch.

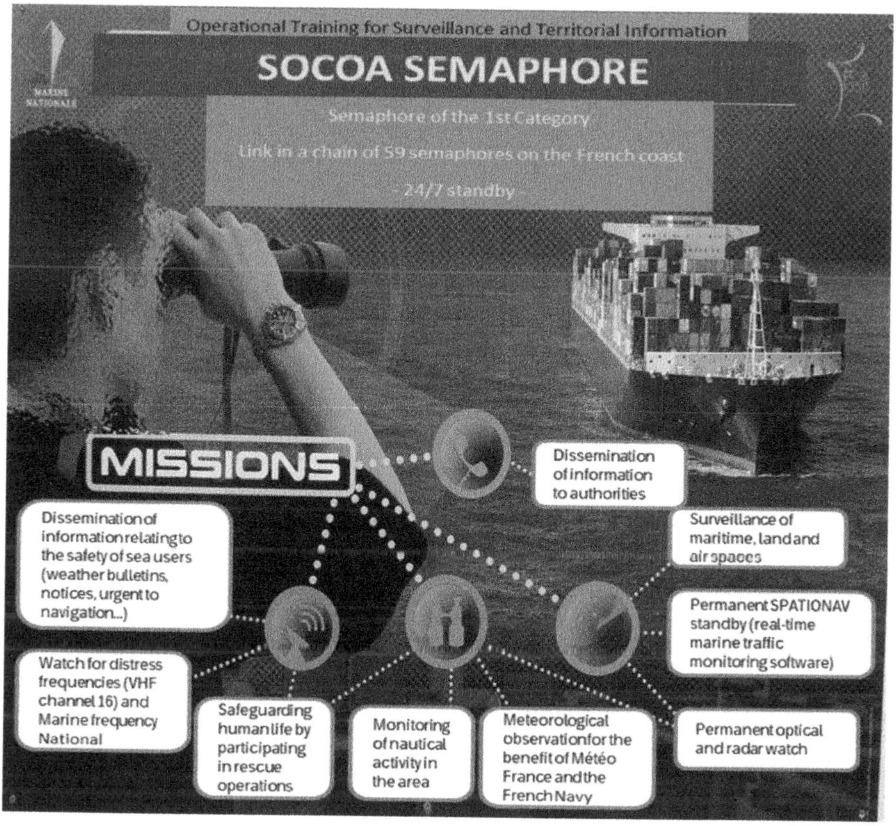

Figure 5.23. *The missions of Socoa semaphore*

A mission, which we can also call a utilization case or use case is defined by (see Figure 5.24) the stakeholder or external system that is the primary recipient of the engagement. There can be several recipients of the mission:

– the initial state of the recipient before the mission;

– the final state of the recipient after the mission;

– the operational context; there may be several;

– external stakeholders or systems that contribute to the mission.

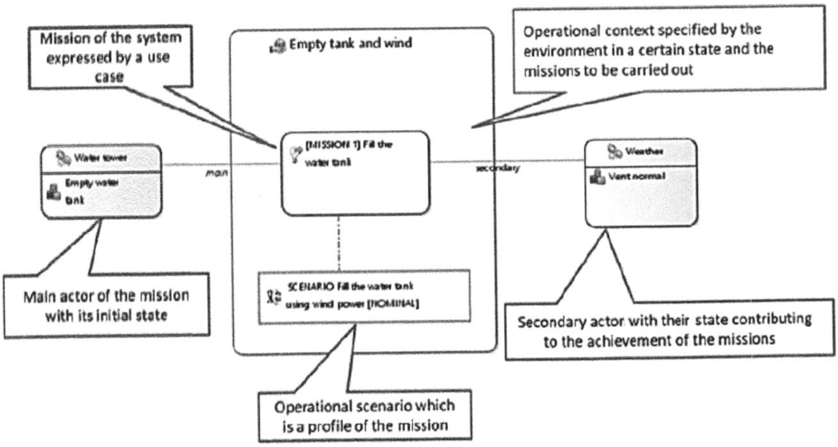

Figure 5.24. *Mission in context*

5.6.2. *Application to the unifying thread example: mission definition*

By going back to the collection of needs, we can identify the uses of the "water pump with renewable energy" system.

Applicant	Capacity	Criterion	Context
Residents	Want the energy produced by the system to be sold back to the grid	?	If the water tower is full and the wind >Vmin m/s
Water tower	Must always be filled	With a water height between Hmin and Hmax	?

Indeed, from these two needs, we can identify the following missions:

– mission 1: fill the water tank;

– mission 2: supply electricity to the grid.

From the life cycle, we can put the missions of the system in context (see Table 5.2).

	Tank		Weather		Mission	
	Full	Vacuum	Wind	No wind	Fill the water tank	Supply electricity to the grid
Context 1		X	X		X	
Context 2		X		X	X	
Context 3	X		X			X
Context 4	X			X		

Table 5.2. *Missions in their contexts*

Mission 1, "fill the water tank", is performed in two operational contexts which are when the tank is empty and there is wind or when the tank is empty and there is no wind.

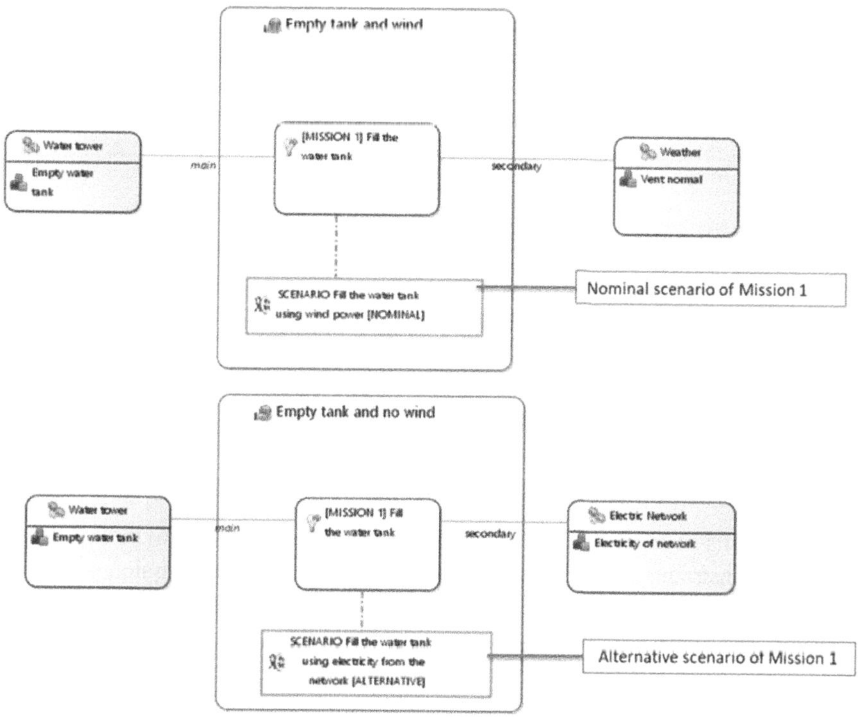

Figure 5.25. *Unifying thread example missions in their contexts*

Figure 5.25 models mission 1 in both operational contexts. We can see the primary actor, which is the beneficiary of the mission, namely the water tower and the secondary actor which is a dimensional contributor, namely the weather. Under each external system, we find the state it is in at the beginning of this mission. The analysis of the output events of the two contexts (see Figure 5.20) does give us the output states at the end of the beneficiary actor's mission, i.e. the water tank is full.

Figure 5.26 represents a mapping of the missions of the "renewable energy water pump" system. If necessary for the analysis of the system requirements, we can break down the missions. This amounts to breaking down goals into sub-goals. This can facilitate the identification of expected services and associated constraints.

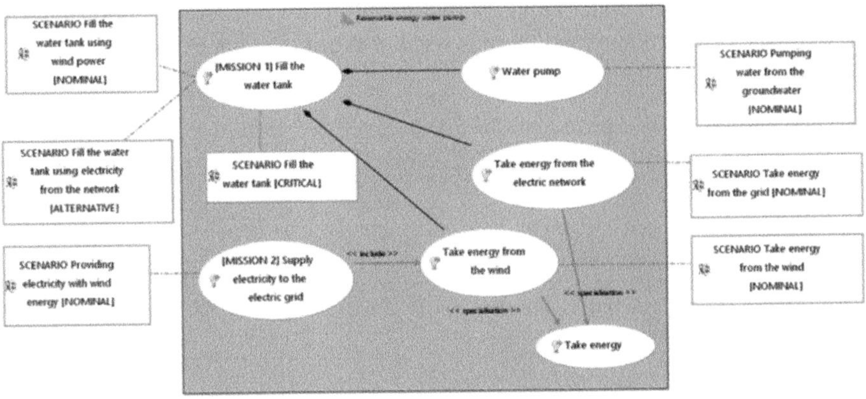

Figure 5.26. *Mission mapping*

5.7. Identification of operational scenarios

5.7.1. *Defining an operational scenario*

To fulfill its missions, a system must provide services to the environment while respecting constraints. An operational scenario corresponds to a mission profile of a system in relation to its environment. It describes the dynamics of the system by listing the exchanges between the environment and the system seen as a black box (AFIS 2004; AFIS 2012). This exchange dynamic between the environment and the system can be modeled by a sequence of actions representing the services rendered or by the exchanges of matter, energy and information representing the transformed flows.

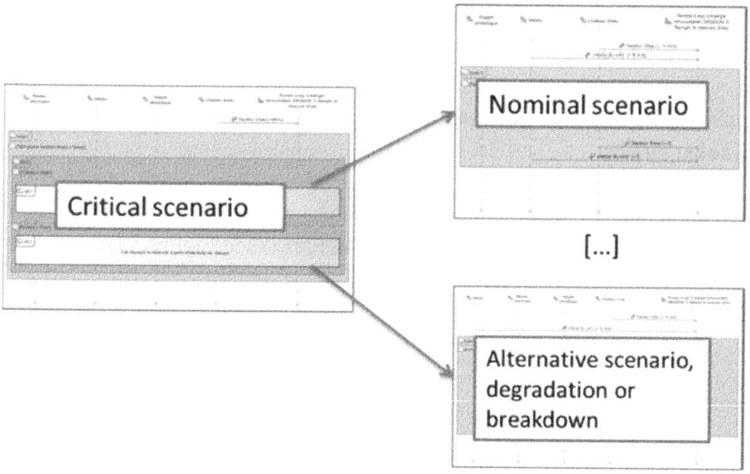

Figure 5.27. *Several types of scenarios*

For the same mission, there can be several types of operational scenarios, namely:

– the **nominal scenario:** this is the execution of the mission when everything goes well. This is the first scenario that must be analyzed;

– the **alternative scenarios:** these are the scenarios that allow the system to achieve the mission goal, but in a different way;

– the **degradation or breakdown scenarios:** these are the scenarios that either allow the mission to be carried out, but with lower performance, or that allow the system to reach a state of failure.

We will call the **critical scenario** the highest level scenario translating the mission with its different dimensioning profiles, i.e. nominal, alternative, degradation or breakdown (see Figure 5.27).

5.7.2. *Application to the unifying thread example: definition of operational scenarios*

For our unifying thread example, mission 1 "fill the tank" can have two profiles namely:

– filling the tank with water using wind power;

– filling the tank with water from mains electricity.

The “fill the water tank with wind power” scenario is the **nominal scenario** of the “fill the tank” mission.

The “fill the reservoir with water from grid electricity” scenario is an **alternative operational scenario**.

We define a critical scenario that allows us to identify the conditions for moving from the nominal scenario to the alternative scenario (see Figure 5.28).

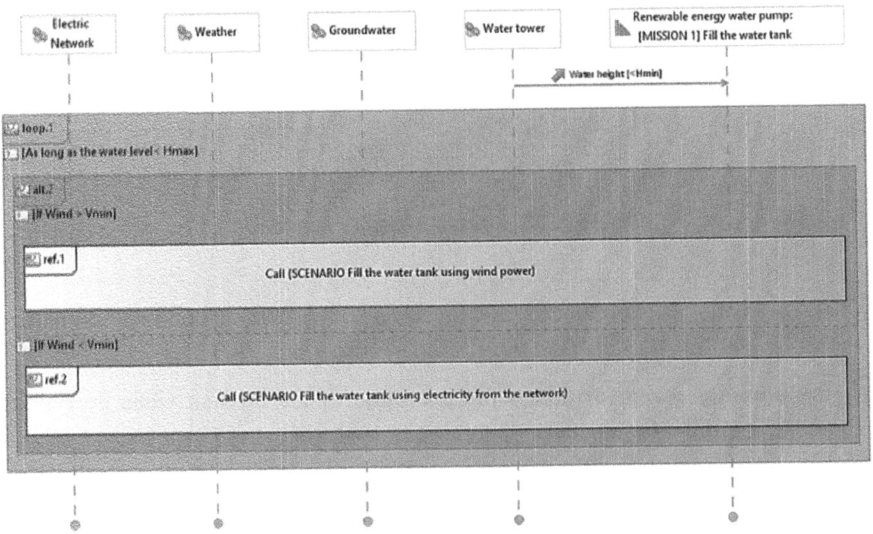

Figure 5.28. *Critical scenario of the unifying thread example*

In the critical scenario, we refer to the nominal scenario and the alternative scenario (INCOSE 2015). In this example, the mission starts if the water level is lower than Hmin; this is the triggering event. Then, the mission lasts as long as the tank does not have a sufficient water height, i.e. lower than Hmax.

To execute its mission, we have two possibilities, the first one corresponds to the nominal scenario if the wind is higher than Vmin. The system uses wind energy as a resource to pressurize groundwater to fill the reservoir (see Figure 5.29).

In the sequence diagram of Figure 5.29, the modeled loop means that the system continues to fill into the tank as long as the water head is less than the maximum height Hmax and as long as there is wind greater than X m/s. At each iteration of the loop, the system measures the water height in the reservoir and the wind speed.

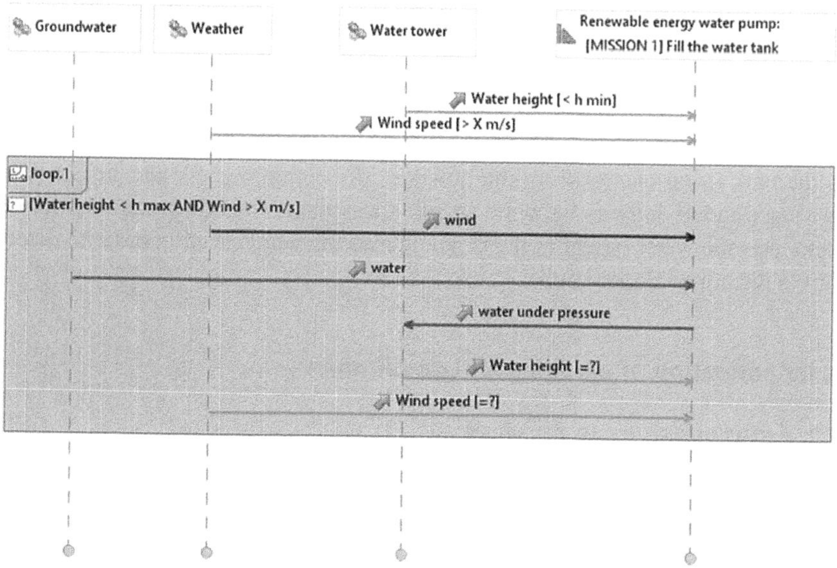

Figure 5.29. *Nominal unifying thread example scenario*

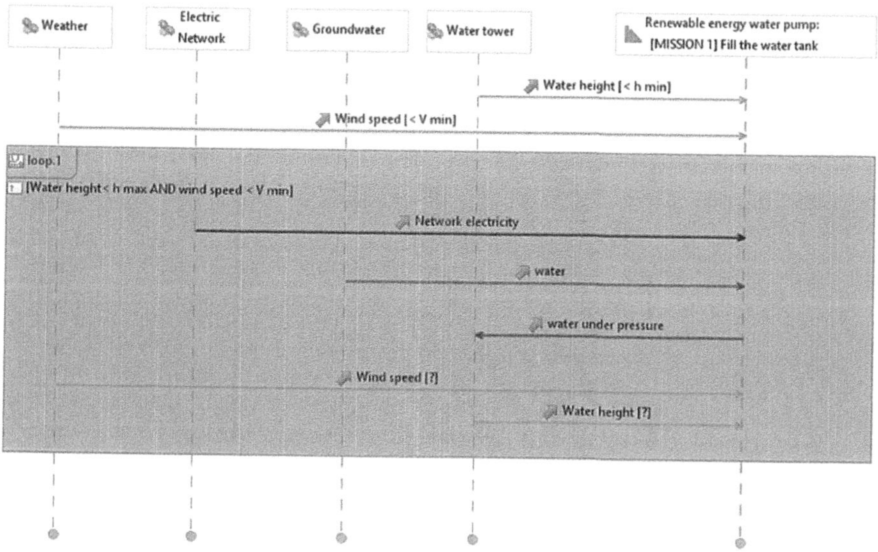

Figure 5.30. *Alternative unifying thread example scenario*

The second is an alternative scenario if the wind is less than Vmin. The system uses the energy of the electrical network as a resource to pressurize the groundwater to fill the tank (see Figure 5.30).

We find in the sequence diagram an iteration loop which means that the system fills the tank using energy from the power grid as long as the water height is less than Hmax and as long as the wind speed is less than Vmin. At each iteration, the system tests the water height in the tank and also the wind speed in order to detect a return to the nominal scenario.

5.8. Identification of services and constraints

5.8.1. *From missions to services*

The services are the expectations of the environment towards the system in terms of use. According to AFIS (2012), a service is a function, capacity or behavior provided by a producer to a consumer. The service is thus seen as an effect expected by the environment with respect to the system's functions. In Micouin (2006), we have a definition of "effect" functions versus "transformation" functions. The first can be the expected service formalized by the user's needs and the second the technical response of the designer, formalized by the system requirements.

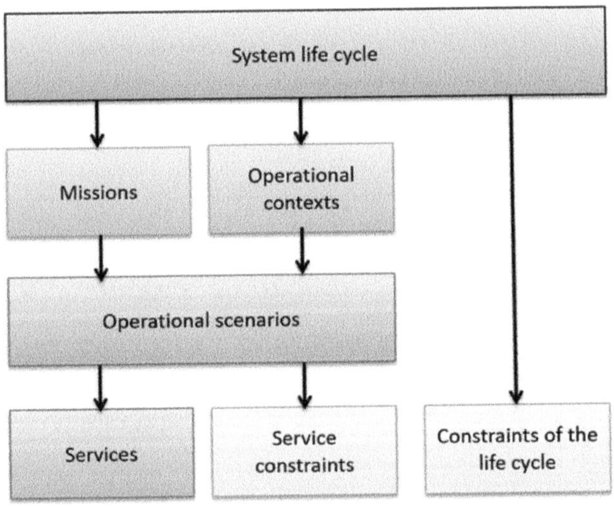

Figure 5.31. *Services and constraints*

The constraints express a limitation that the system must respect. There are several types of constraints (AFIS 2012):

– physical environmental constraints;

– security constraints;

– constraints related to users and the social environment;

– economic, market and competition constraints;

– interface constraints and technological compatibility with the environment;

– regulatory constraints related to the environment;

– logistical constraints.

In Figure 5.31, we have the different relationships between the operational concepts. At the top, we have the life cycle of the system from which we can define the missions and the operational contexts as well as a certain number of constraints (the life cycle constraints). From a mission and an operational context, we can write an operational scenario which is a mission profile. From an operational scenario, we can extract the services and the associated service constraints, also called the quality of service or performance (see Figure 5.32).

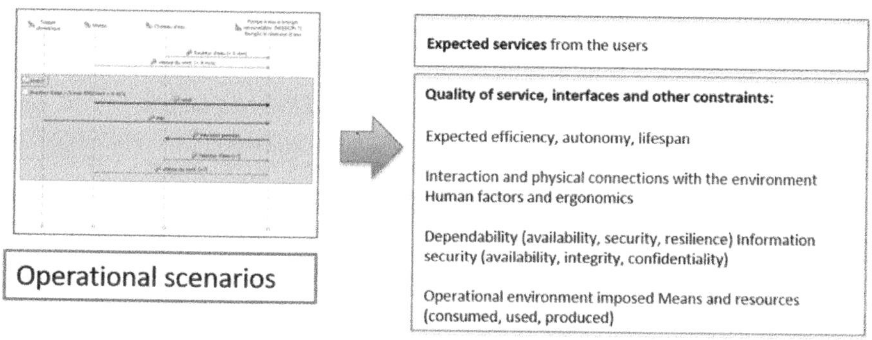

Figure 5.32. *Identifying services and constraints*

5.8.2. *Application to the unifying thread example: identification of services and constraints*

If we analyze the mission of the "fill the tank with water" system through these two operational scenarios, we can identify three services. By analyzing the nominal operational scenario "fill with water with wind power", we can identify the following services:

– use the energy of the wind;

– pump water from the water table.

By analyzing the alternative operational scenario “fill water with grid energy”, we can identify the following additional service:

– use the energy of the electrical network.

To model the system services, we can show them as elementary use cases on the mission map (see Figure 5.33).

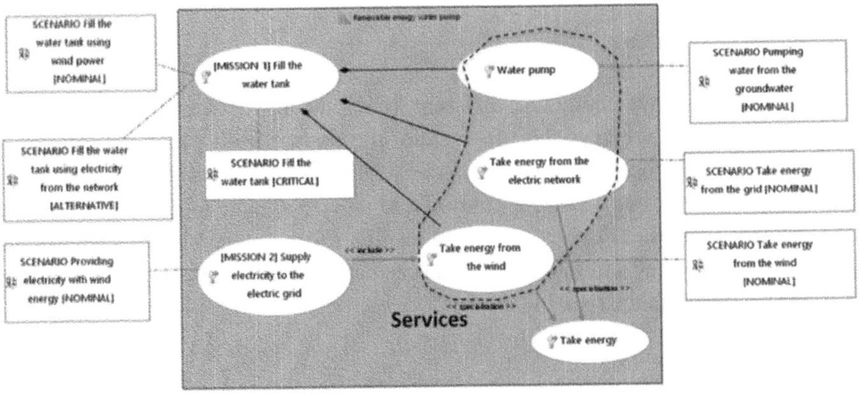

Figure 5.33. *List of services in the unifying thread example*

5.9. Formalization of stakeholder needs

5.9.1. *The input specifications*

The requirements are formalized in the specifications, also referred to as the technical specification of requirements (this name is given for information because it can be the subject of debate despite its existence in certain standards such as MIL-STD 499). It is the responsibility of the project owner. In these input specifications we must find all the elements that allow the project manager to propose a technical response in the form of a specification.

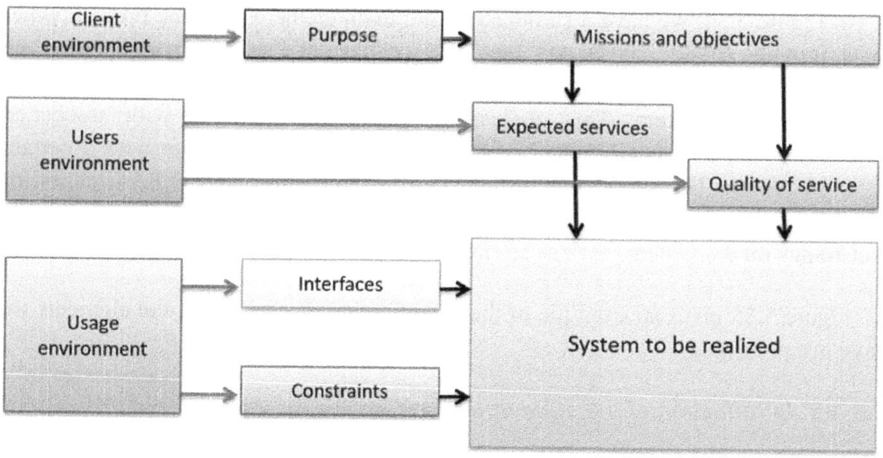

Figure 5.34. *Synthesis of the elements defining the system to be realized (Pollet 2016)*

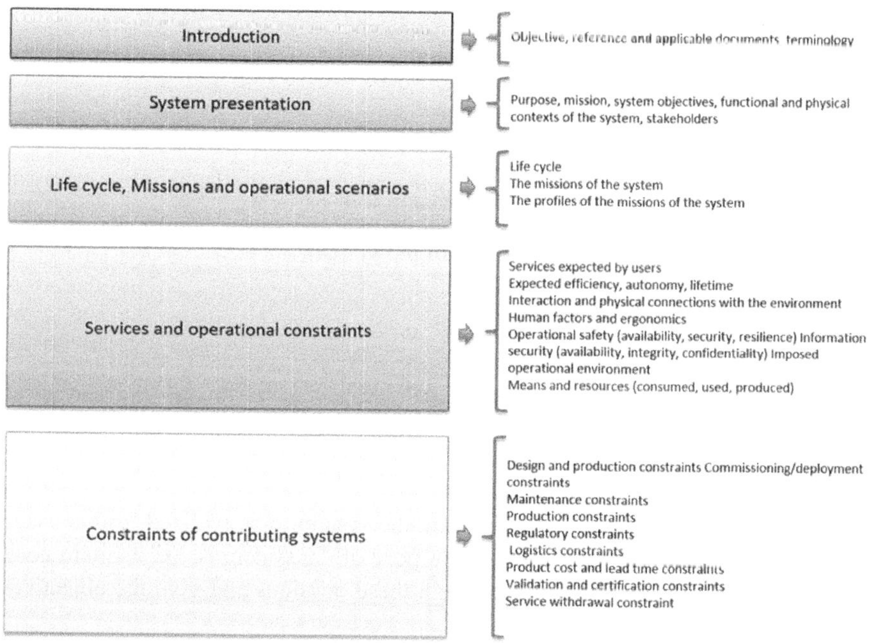

Figure 5.35. *Structure of specification of needs*

In Figure 5.34, we have a summary of the main elements that the project owner must provide (Pollet 2007). On the left, we have the environments where the different demands on the system come from. The client's environment defines the purpose of the system. This goal is translated into missions with associated objectives. The user environment expects services from the system with a certain level of quality. The services with their performance allow the implementation of the missions with their objectives. The user environment imposes interfaces and constraints on the system, such as environmental constraints.

Figure 5.35 gives an example of the specification of needs with the elements we have just presented.

5.9.1.1. *Introduction of the input specifications*

The introduction is standard. We must give:

– the purpose of the document;

– the applicable documents that will be used to draft the specification of needs;

– the reference documents that are used in the specification of needs;

– the terminology with glossaries, acronyms and definitions to assist in reading the document.

5.9.1.2. *Presentation of the system*

The presentation of the system is achieved in four parts, we have:

– the purpose, missions and objectives of the system;

– the functional context;

– the physical context;

– stakeholders.

Purpose, missions and objectives

In this section, the objective is to present the system to be realized from a usage point of view. We have to state what the purpose of the system is, i.e. its main goal. Then, we need to describe this goal through these missions and give the objectives of these missions.

Functional and physical context

The system must be presented in its functional and physical context. For the functional context, the idea is to present the interactions of matter, energy and information flows between the system and its environment. As part of the needs analysis process, it is possible to focus on the desired effect. The transformation part can be completed during the system specification process.

For the physical context, the objective is to resume or complete the interactions, but from a physical point of view, by identifying which interfaces are imposed by the environment.

Stakeholders

The purpose of this section is to present the environment of the system. This presentation is made from the point of view of the stakeholders using the system. This environment will be completed by the stakeholders involved in the realization during the system requirements specification process.

5.9.1.3. *Life cycle, mission and operational scenarios*

The life cycle of the system is presented along with the top-level operational contexts that are generic to all systems.

The operational contexts, at least when the system is in service or in maintenance, can be detailed.

On the in-service contexts, we can identify the missions of the system by presenting at least which stakeholders or external systems are the main actors. This analysis can be extended to other contexts where the system could fulfill missions such as maintenance or others.

Mission behaviors in their operational contexts can be explained through operational scenarios. We will focus primarily on the nominal operational scenarios.

5.9.1.4. *Services and operational constraints*

The operational scenarios show the services that the system must provide. These services will become the functions of the system after the project manager adds the technical principles. In summary, the services must translate the expected effect and the functions, the desired transformation to allow the desired effect.

We must associate the services with the expected operational performance. This will allow the project manager to break them down into functional and global performances on the system.

Among the operational constraints, we have the interfaces imposed by the environment. We also have the human factor constraints that need to be specified. These constraints are important as they influence system performance, interfaces and ergonomics.

Depending on the criticality of the system, the project owner can provide input requirements for operational safety.

The operating environment needs to be clarified. Each of these external systems may have a number of constraints.

Finally, the project owner can impose the resources that the system must use. By doing so, it reduces the technological choices of the project manager.

5.9.1.5. *Constraints of the contributing systems*

Throughout the standard system life cycle, the project owner must define the constraints of the contributing systems.

Design and implementation constraints

The project owner can impose constraints on the way the system is designed, for example, by imposing the application of a standard on the project. On the realization, the owner can impose a supplier to the project manager.

Commissioning/deployment constraints

These constraints may relate, for example, to the training needs of the operators, the number and rate of units delivered of the system that is going to be mass-produced.

Maintenance constraints

The field of maintenance is vast. There may be many requirements imposed by the client, for example, on first-level maintenance, preventive maintenance or planned maintenance.

Production constraints

The client can impose, for example, a number of units produced or the application of a standard on production.

Regulatory constraints

The client may require the contractor to comply with certain regulations. These constraints must be analyzed by the project managers, as they often provide specification elements for the system to be built.

Logistics constraints

The project owner can impose the conditions of transport and storage of the system.

Cost and time constraints on the product

The project owner can give cost and time constraints to the system.

Validation and certification constraints

The project owner can impose constraints to validate the system. It may require compliance with certain regulations that would impose certification requirements.

Withdrawal of service constraint

The project owner can impose constraints for the removal of the system from service such as dismantling and recycling.

5.9.2. *Application to the unifying thread example: formalization of needs*

5.9.2.1. *Presentation of the system*

For the presentation of the system, we need to create an introduction integrating the purpose of the system with the main mission it has to perform and quantifiable objectives. For our unifying thread example, this could be:

The municipality wants a new pump system that will fill the reservoir with groundwater up to a height of Hmax as soon as the height is lower than Hmin and preferably using renewable energy

Generally, the introduction of the system is completed by a presentation of the functional and constructional contexts. The **functional context** is a map representing the main MEI flows exchanged between the environment and the system. The **constructional context** gives the interfaces imposed by the project owner on the system.

5.9.2.2. *Expected services and quality of services*

For example, for the mission "fill the tank", we can formalize the following needs.

Need

The tank must be able to fill with water from the water table up to a height of 95% of Hmax, when the water height is lower than Hmin in the context "empty tank and wind" or "empty tank and no wind".

Need

The water tower tank must be filled from the minimum height to 95% of the maximum height of water within 36 hours in an "empty tank and wind" or "empty tank and no wind" operational environment.

When there is wind, we have the following **environmental constraint** which is also an assumption.

Constraint

The weather forecast calls for winds between 0 m/s and 8 m/s when the system is in operation.

When there is no wind, we have the following **assumption**.

Assumption

The electrical grid must be able to supply 220 vac at 50 Hz to the system when the tank is empty and there is no wind.

For the mission "supply electricity to the electrical grid", we can formalize the following requirement.

Need

The electrical grid must be able to receive electrical power from the system at 220 vac and 50 Hz when the tank is full and there is wind.

5.9.2.3. *Constraints of the contributing systems*

We have illustrated the different categories of constraints with an example applied to our unifying thread example.

Design and implementation constraints

The wind turbine supplier requires a wind turbine operating with winds between 3 m/s and 8 m/s.

Commissioning/deployment constraints

The EDF network requires that the system's operation does not disrupt its electrical network in the commune for more than two hours.

Maintenance constraints

The water system operator requires that maintenance can be done without removing the wind generator.

Regulatory constraints

The DGAC requires that the system be equipped with a visual device to signal to light aviation.

Logistics constraints

The DDE imposes that the system during transport does not exceed an authorized weight of 36 tons.

Cost and time constraints on the product

The municipality requires that the system be delivered and installed within three years of signing the contract.

The municipality requires that the sale of electricity compensates for the purchase of electricity and a 2% reduction in the water bills of the citizens.

Validation and certification constraints

Europe requires that the system meets the current environmental standards.

Withdrawal of service constraint

The municipality requires that the system be over 90% recyclable.

5.10. Conclusion on needs analysis

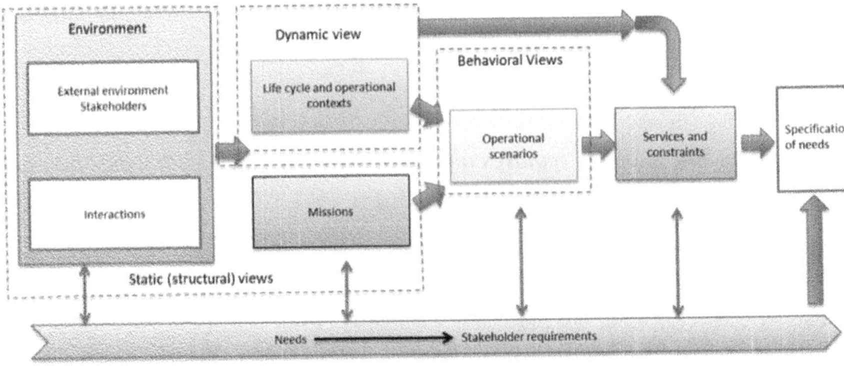

Figure 5.36. *Summary of needs analysis*

The key points of the needs analysis are summarized in Figure 5.36. The starting point is the identification of the system environment. The analysis of the interactions makes it possible to define the limits of the system as well as those of the direct environment.

The purpose of the system is expressed through its missions. The interactions and missions form the static and structural vision of the system in its environment.

The dynamic vision of the environment is built with the life cycle and operational contexts of the system.

From the missions and operational contexts, we can write the operational scenarios of the system.

From these scenarios, it is easy to deduce and formalize, in the form of needs or stakeholder requirements, the services and associated service constraints.

All of these stakeholder requirements are prioritized in a set of specifications.

6

Requirements Specification

6.1. Objective of the requirements specification process

A requirements repository is a structured collection of requirements that can be managed in configuration. For the development of a system, the project manages two requirement repositories, namely (AFIS 2012):

– the initial requirements repository which expresses the needs of the stakeholders interested in using the system;

– the system requirements repository which refers to the initial requirements by adding the constraints of the selected system concept. This repository must be validated by the client.

The collection of stakeholder requirements can be put into a database or into a document that is called a technical specification of need or **specifications**.

The collection of the system's technical requirements can be put into a database or into a document which is called a technical specification of the system or **system specification**.

REMARK.– If the requirements are put in a database, it is possible to extract a document from it.

The objective of this process is to define a repository of requirements for the system to be made (see Figure 6.1).

For a color version of all the figures in this chapter, see www.iste.co.uk/hygounenc/system.zip.

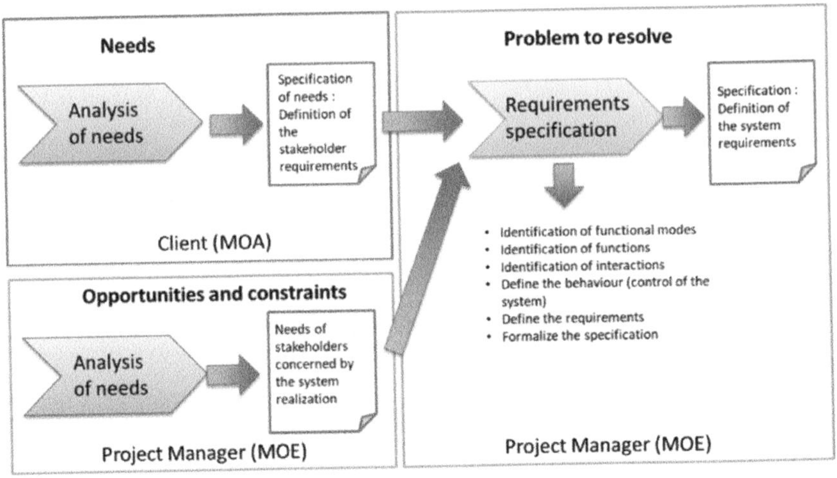

Figure 6.1. *Inputs and outputs of the needs analysis and requirements specification*

At the input of this process, we have the outputs of the needs analysis of the stakeholders interested in the use and the stakeholders interested in the realization of the system. At the end, we will have a collection of requirements in the form of a specification, and the verification and validation plan of the system (see Figure 6.2).

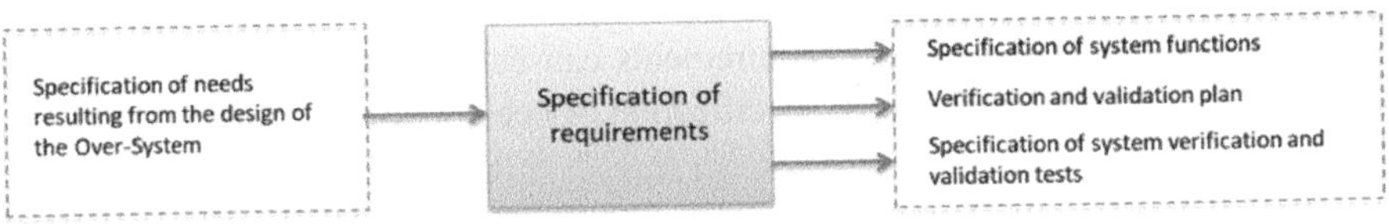

Figure 6.2. *SIPOC system specification*

The requirements of this repository shall:

– be validatable, i.e. they must meet the needs of the stakeholders;

– be verifiable, i.e. testable, on the integrated system with one of the following methods: analysis, modeling, simulation, testing, similarity;

– enable design, i.e. be feasible and doable. The system requirements should be able to be allocated to the system components (subsystems).

6.2. Identification of the system's functional modes

6.2.1. *Definition and purpose of the functional modes*

A **functional mode** is a functional state of the system representing a period of time during which the system has a coherent behavior responding to missions for one or more operational contexts.

During this functional mode, the system activates and deactivates one or more functions. The entry in a functional mode is done after the appearance of an event that is external or internal to the system.

An **event** can be a change of state of an external system or an MEI (matter, energy, information) flow with a certain value or a combination of flow and state.

The functional modes allow us to study the system from a temporal perspective. In the specification of the system, the level of detail should correspond to the coverage of the operational contexts in response to the different missions. This discrete time study controls the complexity of the system.

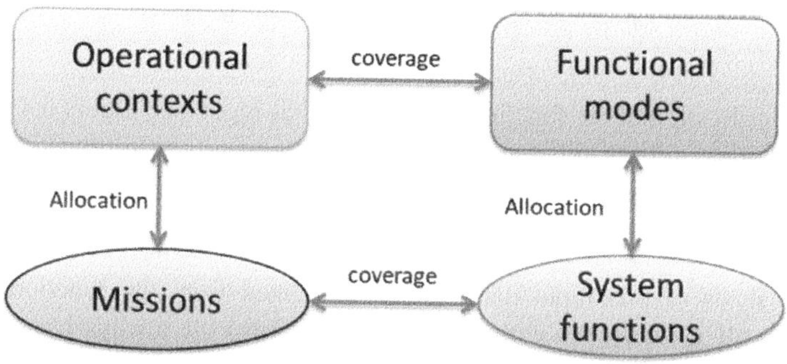

Figure 6.3. *Magic square*

At the specification level, the level of detail must be in line with the operational vision. The balance between the operational and functional vision of the specification level forms a "magic" square. The missions of the system are in operational contexts that are covered by functional modes, which provide functions that cover the missions (see Figure 6.3).

6.2.2. *What to do to identify functional modes*

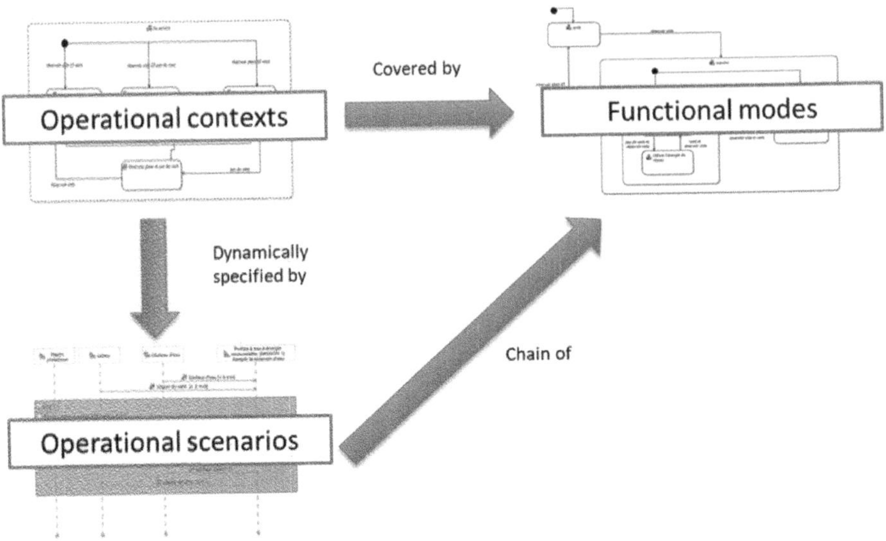

Figure 6.4. *Inputs to identify operating modes*

The identification of the functional modes used as input (see Figure 6.4):

– operational contexts;

– the missions put in the form of operational scenarios, i.e. the mission profiles.

The analysis of this input data can be initialized from the functional modes common to all systems. As with the life cycle, where we always have the same first-level operational contexts, a system has, in the vast majority of cases, three high-level functional modes, with energy, namely (Duysinx 2016):

– working mode: system operating procedures;

– shutdown mode: system shutdown procedures;

– failure mode: degraded system operating procedures.

To these three modes, we can also add the "stopped" mode where the system is without energy (see Figure 6.5). We can also add the system initialization mode which can be integrated with the working mode.

The analysis of the operational contexts allows us to identify the functional state of the system in order to adapt to the state of the environment and to respond to the requests, i.e. the services that allow the missions to be fulfilled.

The analysis of operational scenarios allows us to identify changes in the system's functional mode. Indeed, at the beginning of the scenario, the system is in an initial mode.

When the initiating flow is received, the system goes into a functional mode that allows the activation of the functions that implement the expected services.

At the end of the scenario, the system can switch to another functional mode.

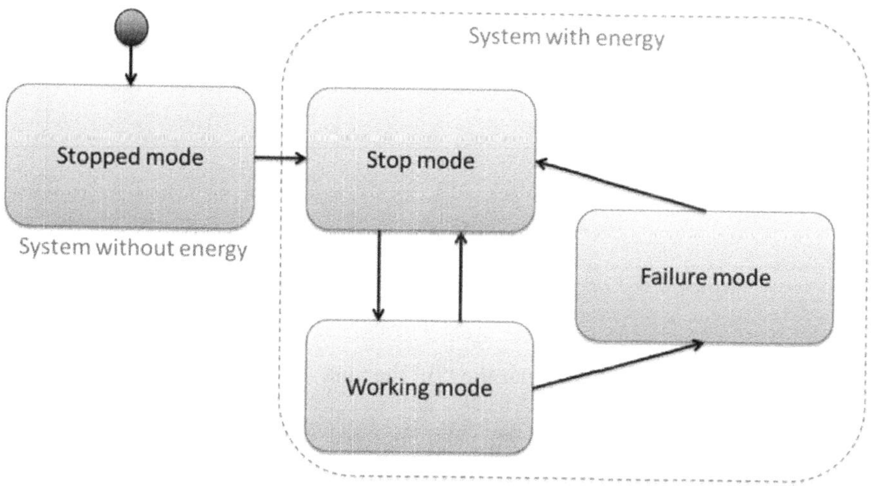

Figure 6.5. *Functional modes of a system*

6.2.3. *Application to the unifying thread example: identifying the functional modes*

For our unifying thread example, we will consider two functional modes as a starting point for our analysis, namely On and Off.

The analysis of the operational contexts and scenarios allows us to identify three operating procedures, which will give the functional modes:

– fill the tank and use the wind energy;

– fill the tank and use the energy of the network;

– provide energy to the grid.

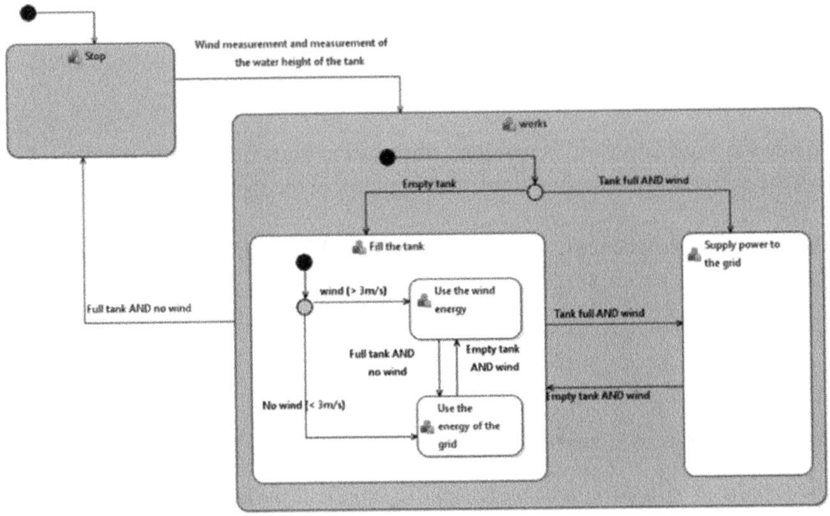

Figure 6.6. *Functional mode of the unifying thread example*

We can group the first two under the "fill the tank" functional mode.

To finalize the analysis, we put the functional modes in the form of a state machine by identifying the events that allow us to switch from one mode to another (see Figure 6.6).

In the unifying thread example, we did not treat the failure modes and simplified the shutdown mode as a single mode with or without energy.

6.4. Identification of system functions

6.4.1. *What is a system function?*

To meet the missions, users expect services that are implemented by the system through its functions. In Figure 6.7, we take our example on Socoa semaphore. The pictograms on the poster represent the functions provided by the semaphore to enable the Navy to carry out its various missions.

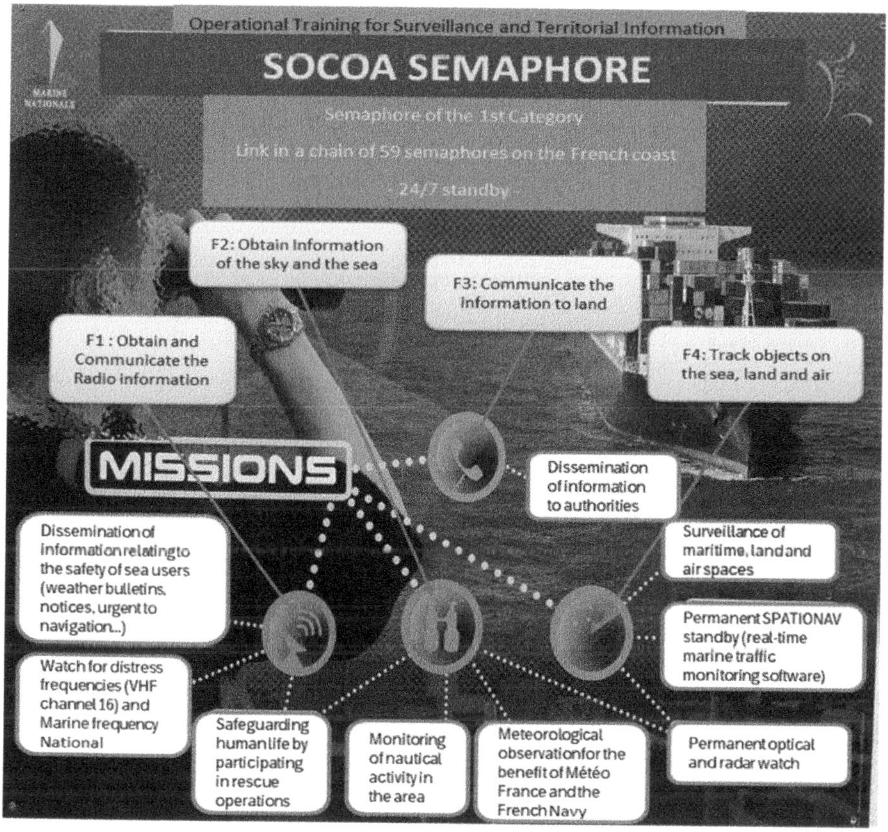

Figure 6.7. *The functions of Socoa semaphore*

To identify the functions, we can:

– analyze the expected services from the project owner’s point of view and complete or translate them according to the project manager’s point of view;

– analyze operational scenarios that describe the system’s missions through the exchange of MEI flows between the system and the environment;

– analyze the functional chains of operation and control that have been identified during the analysis of interactions with the environment and translate them into functions.

The designer is free to choose to implement one or more of these approaches. They are complementary and should lead to more or less the same result.

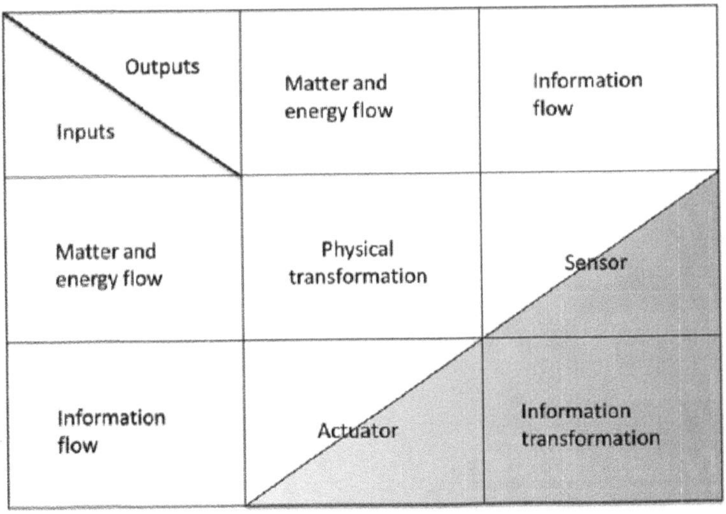

Figure 6.8. *Transformation matrix (Pollet 2016)*

A function is an abstract entity that transforms the incoming MEI flows into outgoing MEI flows (AFIS 2012). Due to the abstract nature of the function, all types of inputs and outputs are possible (see Figure 6.8).

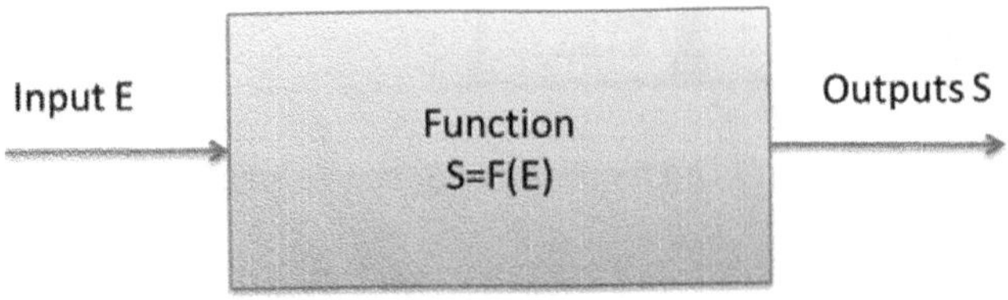

Figure 6.9. *Function S = F(E)*

A function links its inputs to its outputs through the relation S = F(E) (see Figure 6.9). In its simplest form, it can be associated with three basic changes, each of which brings value:

– the **time** change **T**;

– the change of **space E**;

– the change of **shape F**.

By associating the TEF changes with the MEI flows, we obtain the MEI–TEF matrix (Meinadier 1998; Faisandier 2013) (see Figure 6.10).

	M Matter	E energy	I information
T time	storage	accumulation	memorization
E space	transport	transport	communication
F form	transformation	transformation	treatment

Figure 6.10. *MEI–TEF matrix*

To remain consistent with the combination of flows (see Figure 6.8), we must add to the basic "transformation" functions the "transmutation" functions to cover the cases of sensors and actuators (Meinadier 1998). Indeed, a sensor changes the nature of a flow, for example, from an energy flow to an information flow. For the actuators, when we want to translate the **expected effect**, we have an information set point in input which gives in output an energy flow, for example, a translation. If we are looking for completeness by modeling the real transformation, then we will have to identify the matter or energy flow at the input of the function in addition to the set point.

6.4.2. *What to do to identify the system functions*

Analysis of expected services

From the list of services identified in the specification of needs, the designer reformulates them by adding the selected system concepts.

Indeed, the identification of services and functions is a question of perspective. In the requirements analysis, we look for the services that will be realized by the system functions and in the requirements specification we look for the functions that realize the services (see Figure 6.11). In the literature, we speak of identifying **use functions**. We have deliberately separated the two concepts in order not to remove consistency in the two identification activities and introduce ambiguity. Indeed, from a project owner's point of view, the user will express the service function as an expected effect following an input action. We can speak of an "effect" function (Micouin 2006). For example, the driver transmits an acceleration command and the car accelerates. From the project manager's point of view, this service function requested by the project owner must be repeated in order to show the concepts that have been retained for the system. This technical recovery gives an operation function. For example, if the project manager proposes an electric car, the function transforms electrical energy into kinetic energy by following an acceleration set point.

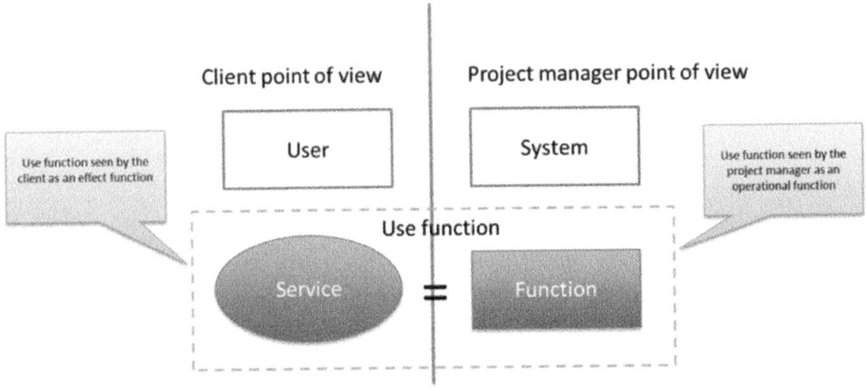

Figure 6.11. *Service and function, a question of perspective*

Operational scenario analysis

By analyzing the operational scenarios, we can identify the system functions (see Figure 6.12). Indeed, one or more functions are hidden behind the MEI flows entering and leaving the system. These functions must reflect the transformation or the transformation chain.

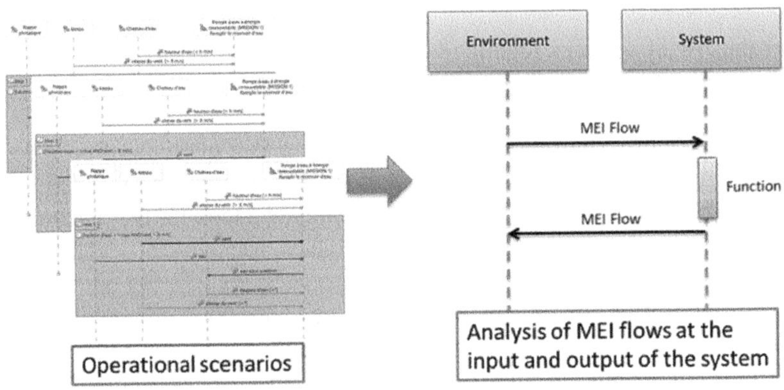

Figure 6.12. *Identifying functions in operational scenarios*

Functional chain analysis

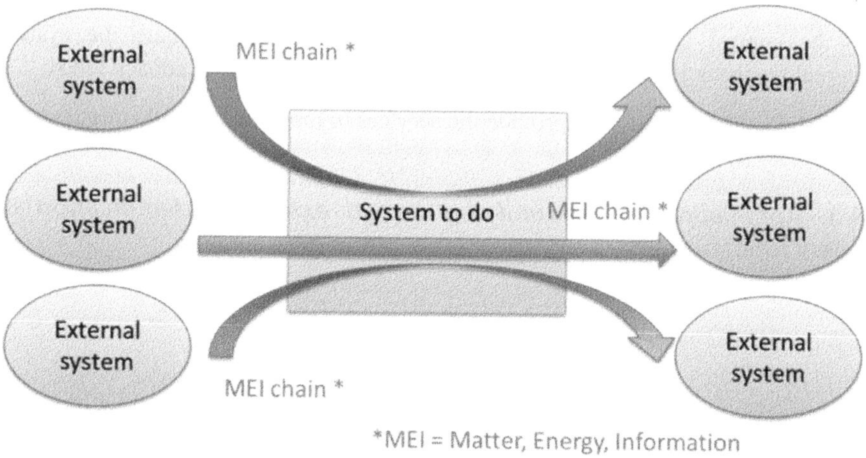

Figure 6.13. *Functional chains that run through the system*

The analysis of the block diagram of the interactions of matter, energy and information (MEI) with the environment makes it possible to identify the functional chains which cross the system (see Figure 6.13).

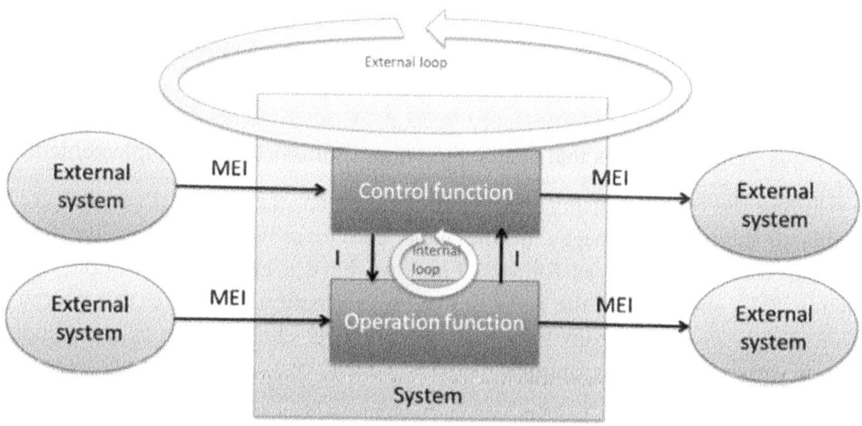

Figure 6.14. *Control and operation chain*

The functions of the system implement one or more functional chains. We have two types of functional chains (see Figure 6.14):

– the control chains which allow the environment to regulate the system (external feedback loop) or the system to regulate itself (internal feedback loop);

– the operating chains that provide the services to meet the system's missions.

6.4.3. *Application to the unifying thread example: identification of system functions*

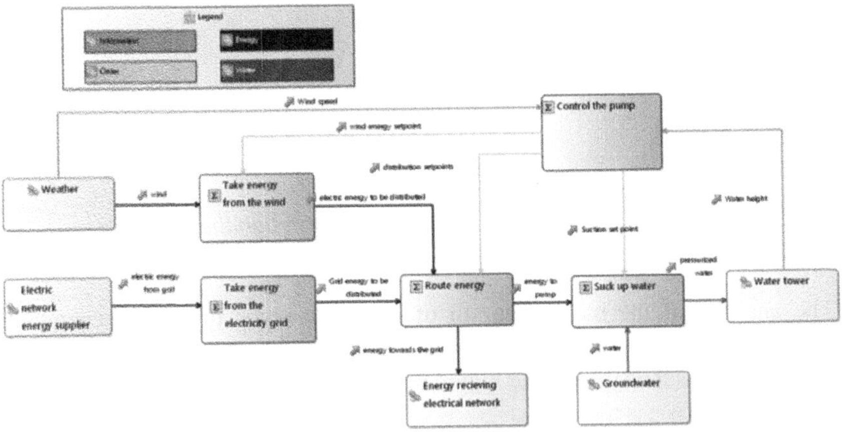

Figure 6.15. *Functions of the unifying thread example system*

We analyze the services that enable the system's missions to be implemented:

– Filling the reservoir using the power of the wind: the expected services are to be able to capture the energy of the wind and to provide a flow. The system functions we identify are as follows. First, we have a chain of operation with a function "take energy from the air" followed by a function that routes this energy to a function "suck water from the water table". This chain is driven by a global function which is "drive the pump".

– Filling the reservoir when there is no wind: the expected services are to be able to capture energy from the electrical grid and to provide a flow. The system functions we identify are as follows. We have a new chain of operation with a function "take energy from the electrical network" followed by a function that will route this energy to the function "suck water". The whole thing is driven by a global function that is "drive the pump".

– To supply electricity to the grid when the reservoir is full and there is wind: the expected services are to capture the wind energy and transmit this energy to the electrical grid. The system functions we identify are as follows. We have a third chain of operation with the functions "take the energy from the air" followed by the function that will route this energy to the electrical network.

6.5. Identification of external interactions

The objective of this activity is to consolidate the identification and characterization of:

– external functional interfaces namely the flows of matter, energy and information (MEI) exchanged with the environment (see Figure 6.16);

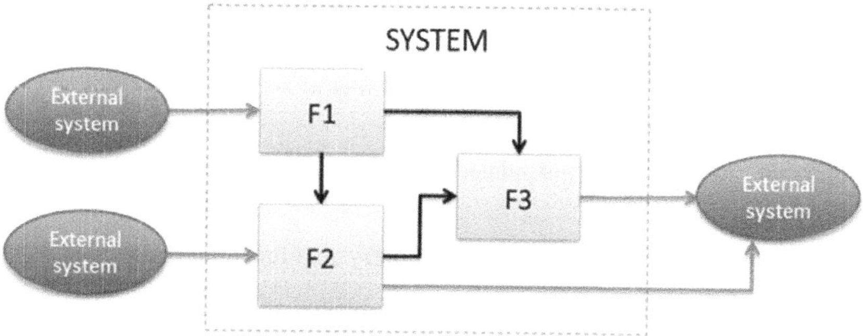

Figure 6.16. *Functional interfaces*

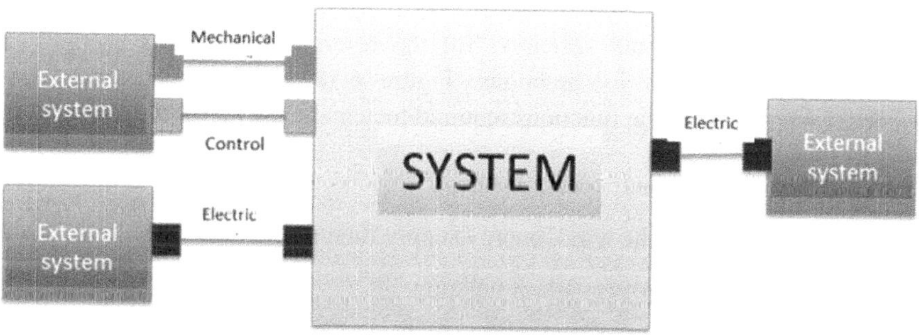

Figure 6.17. *Physical interface*

– external physical interfaces to know the links that exist between the system and the environment. These links can be, for example, mechanical or electrical and are the vectors of the MEI flows (see Figure 6.17).

As an input to this activity, we have the map of external interactions from the needs analysis, the operational contexts and the functional modes of the system.

6.6. Defining system behaviors

6.6.1. *Objective of defining the functional behavior of the system*

Defining system behavior means studying its dynamics during and between operating modes and how it is controlled or regulated.

With the functions and modes of the system defined, the objective of this activity is to rework the operational scenarios by replacing the system with its functions. This will allow:

– functions and modes to be refined;

– external flows to be consolidated;

– internal flows to be consolidated or identified (i.e. flows between system functions);

– operational performance to be allocated to system functions and modes;

– the control and regulation of the system to be studied.

6.6.2. *Application to the unifying thread example: definition of a functional scenario*

We take the operational scenario "fill the reservoir with wind energy" and replace the system with its functions. Figure 6.18 presents the system level functional scenario with the functions required to achieve the mission.

This scenario implements four functions:

– taking energy from the wind: energy supply function;

– controlling the pump: regulation function;

– routing energy: power distribution function;

– sucking up water: energy conversion function.

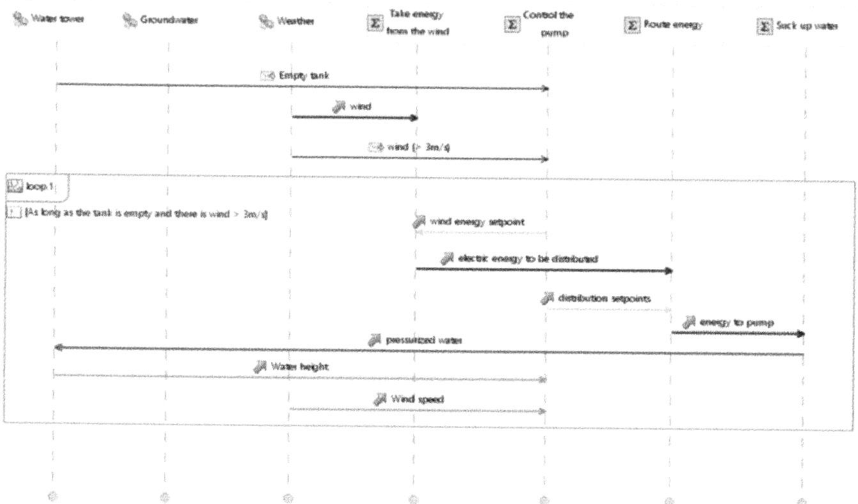

Figure 6.18. *Functional scenario of the unifying thread example*

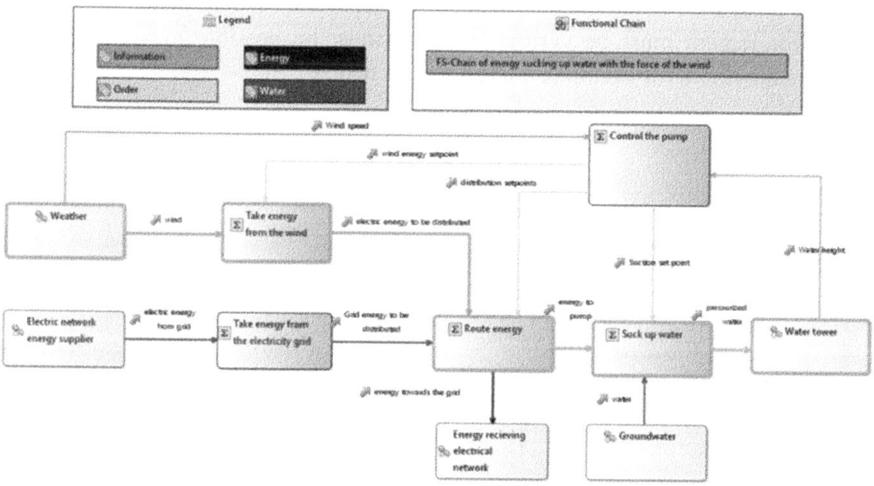

Figure 6.19. *A chain of operations on the unifying thread example*

This functional scenario shows two functional chains:

– a control chain: in the first part of the scenario, we have the "control the pump" function which regulates the system by transmitting instructions to the "take energy from the wind" and "suck up water" functions based on environmental information such as wind speed and water height in the tank;

– a chain of operations: the second part of the scenario models the energy transformation chain to pressurize the groundwater. This functional chain of operation is shown in Figure 6.19.

6.7. Defining the system requirements

6.7.1. *Why define the system requirements?*

DEFINITION.– A system requirement is a condition or capability that the product or a component of the product must possess to fulfill a contract, comply with a standard, specification or other formally imposed document (Badreau 2014).

The system requirements are properties expected on the integrated system. They must meet three criteria:

– a requirement must meet a need, i.e. it must partially or totally cover a need in the environment;

– a requirement must be verifiable, i.e. it must be possible to verify it by a verification method, for example, a test on a test bench;

– a requirement must enable the design of the system, i.e. it must be feasible.

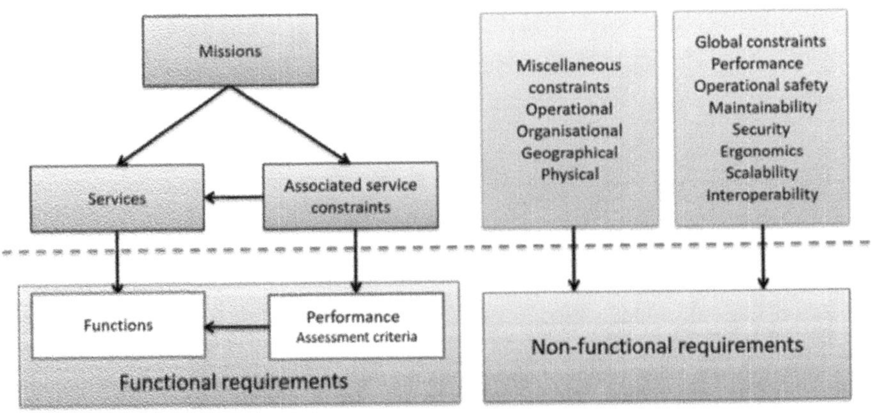

Figure 6.20. *Definition of requirements (Meinadier 1998)*

The requirements of a system are separated into two families, namely the functional requirements and the others, which we will call the non-functional requirements (Figures 6.20 and 6.21).

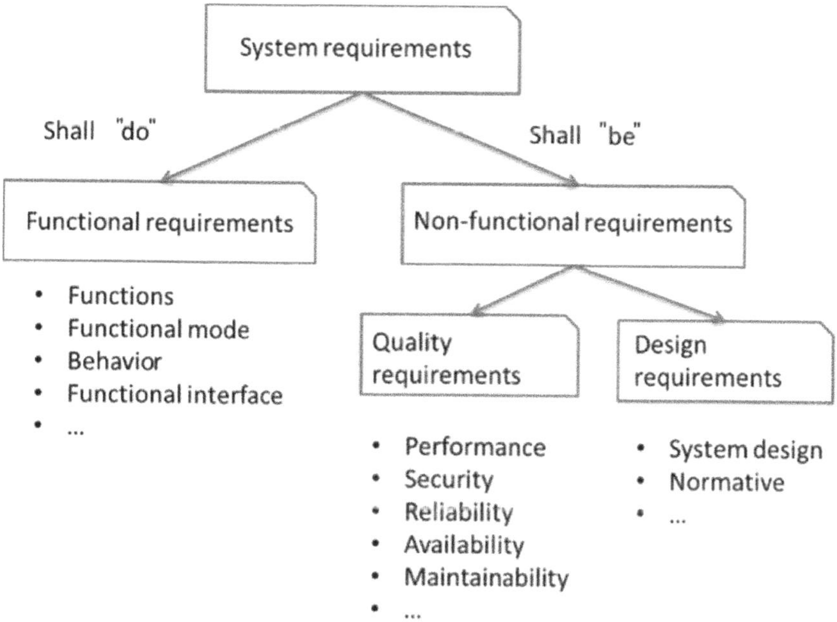

Figure 6.21. *Requirement category (Badreau 2014)*

The system requirements must meet needs, which are derived from the needs analysis process and are formalized in the specifications. These needs must be complemented by the needs of the stakeholders interested in the realization of the system. The requirements are composed of the system's missions and a set of constraints. Missions are implemented by expected services with associated service constraints. Services are the "effect" functions provided by the system to the environment. After analysis, the services are broken down and refined into "transformation" functions, which are the functions of the system and will have to be implemented through the system components. The service constraints are broken down into assessment criteria associated with the system functions. The functions and associated criteria are formalized in **functional requirements**.

DEFINITION.– A **functional requirement** is a system property that specifies what the system must do under what condition and with what performance criteria. According to AFIS (2012), it prescribes an action, a behavior, an expected result or a service provided by the system.

The constraints of the user stakeholders are translated into **non-functional requirements**. In the same way, the constraints of the stakeholders involved in the realization of the system are also translated into non-functional requirements.

DEFINITION.– A **non-functional requirement** prescribes what the system should be. This can be, for example, a global capability expected from the system such as availability, maintainability, scalability, security, interoperability or ergonomics.

6.7.2. *What to do to define the system requirements*

To be able to formalize the **functional requirements**, we must have as input:

– system functions;

– the MEI flows exchanged (material, energy, information);

– functional modes;

– the events;

– functional scenarios.

With this information, which must respond to the missions and services in the operational contexts that come from the needs analysis, we can use the canonical form (from Figure 6.22) to write the functional requirements (Krob 2017).

Applicant	Ability	Criteria	Condition
The "system component"	Shall "do something"	With "a target level of performance"	In a given context

Figure 6.22. *Functional requirement template (Krob 2017)*

For Figure 6.22's template, the requester is the system, a component of the system or a function of the system, the capability is the desired action or effect, the criterion is the quantifiable element that measures the expected action or effect, and the condition is the context in which we wish to have this property. For each function of the system, we will study its static and dynamic environment and try to answer these questions (Figure 6.23):

– What are the inputs to the function? Determine the input assumptions.

– What are the outputs of the function? Determine the expected effect.

– When is the function activated? Determine the functional mode(s) where the function is used.

– What are the inputs that lead to this activation? Determine who drives the function (event, control flow, information flow).

– What does the function transform? Determine the action verb.

From the response elements, one or more requirements can be written using the functional canonical form. Some of these elements can also be used to draft performance requirements.

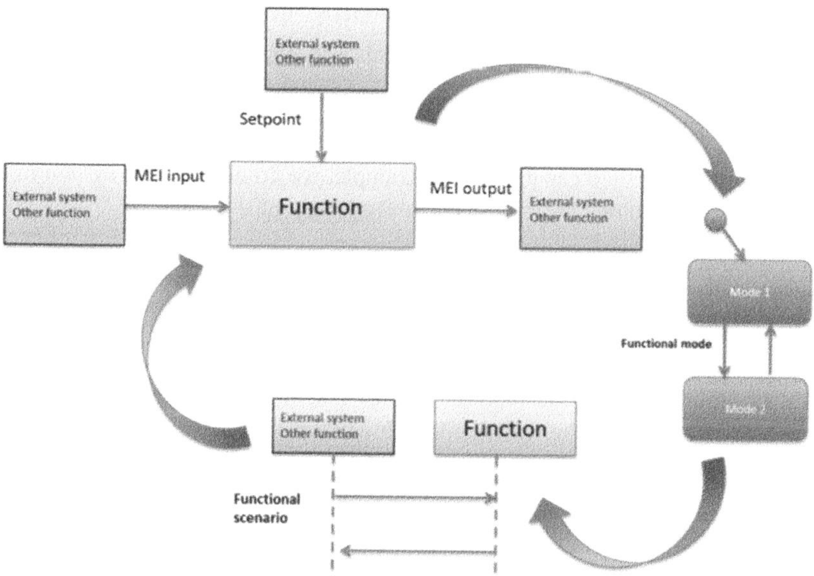

Figure 6.23. *Specification of a function*

For non-functional requirements, we must have as input:

– the constraints of the stakeholders using the system;

– the constraints of the stakeholders involved in the realization of the system;

– the life cycle of the system.

With these elements, we can formalize the non-functional requirements using the organic requirements template (Figure 6.24) proposed by Krob (2017).

Applicant	Ability	Criteria	Condition
The "system component"	Shall "BE made up of something"	With "a target level of performance"	In a given context

Figure 6.24. *Template of organic requirements (Krob 2017)*

On the template of Figure 6.24, the requester is the system or a component of the system, the capability is the concrete property desired, the criterion is the quantifiable element that checks this property and the condition is the context in which we want to have this property.

REMARK.– For this template, the "must be made of something" capability must be matched to the specified capability. For example, we do not say "the system must be made of a blue color", but "the system must have a blue color".

6.7.3. *How to define system requirements: application to the unifying thread example*

We are going to write the requirements of the functional chain that allows for covering the mission "fill the water tank with wind energy". This functional chain of operation is composed of three functions (see Figure 6.19) namely:

– taking energy from the wind;

– routing energy;

– sucking up water.

This functional chain is activated in the functional mode "fill the tank and use the wind energy".

Internal behavioral requirements

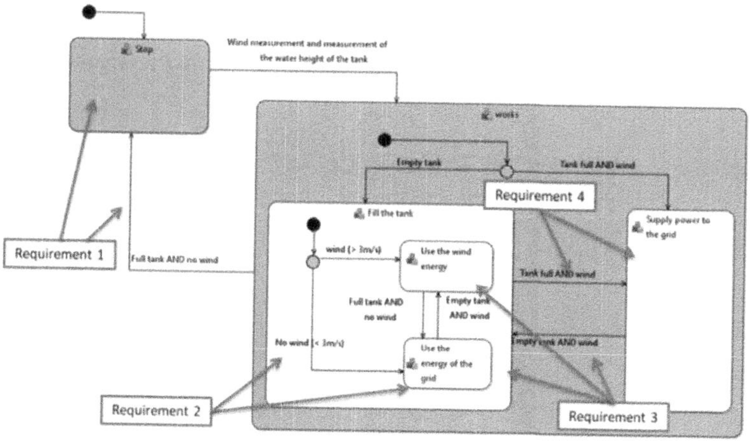

Figure 6.25. *Requirements for functional modes*

For the state machine of the functional modes, here is a requirement example allowing the properties of the internal behavior of the system to be specified according to the definition of Krob (2014) (see Figure 6.25).

System requirement 1

The system should go into off mode if the tank is higher than Hmin and the wind speed is lower than Vmin from the on mode.

System requirement 2

The system shall switch to the "fill the tank and use grid power" mode if the water height is less than or equal to Hmin and the wind speed is less than Vmin.

System requirement 3

The system shall switch to "fill tank and use wind energy" mode if the wind is greater than or equal to Vmin and the water height is less than or equal to Hmin from the off modes or provide energy to the grid or fill the tank and use grid energy.

System requirement 4

The system shall switch to "generate electrical energy for the grid" mode if the water height is between Hmin and Hmax and the air velocity is greater than or equal to Vmin from the fill the tank mode.

Requirements on the function "take energy from the wind"

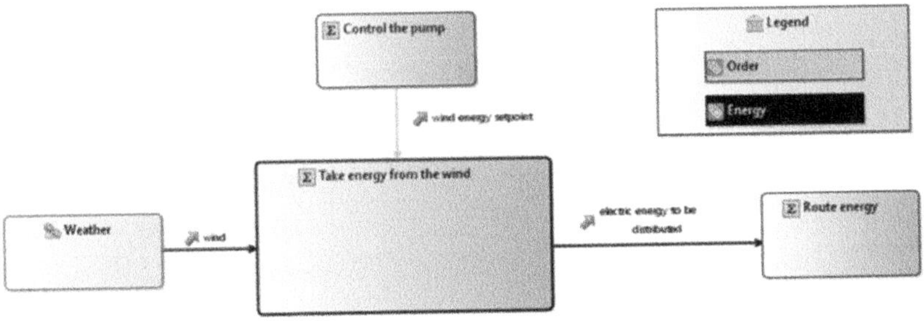

Figure 6.26. *Unifying thread example take energy from the air function*

The function "take energy from the wind" takes as input the force of the wind and as output the electrical energy to be distributed. As the use of a wind generator is imposed by the specifications, we can directly write two system requirements translating the two stages of energy transformation (see Figure 6.26).

System requirement

The system must transform the wind speed into mechanical power between **Pmin** and **Pmax** if the wind is between **Vmin** and **Vmax** and if the control setpoint is "wind energy" in the "suck water with wind energy" operating mode.

System requirement

The system must produce electrical energy between Pelecmin and Pelecmax from the mechanical energy of the wind Pmin and Pmax in "fill the tank and use the wind energy" or "produce electricity for the grid" mode.

NOTE ON THE USE OF VARIABLES IN REQUIREMENTS.– For example, instead of:

The car must have a maximum travel speed of 130 km/h on the highway.

We can write:

The car must have a maximum travel speed equal to Vmax on the highway.

With *Vmax = 130 km/h*, which can be put at the end of the document or in a separate document.

The advantages and disadvantages of this approach are listed in Table 6.1.

Advantages	Disadvantages
Facilitates the evolution of the document	Loss of readability of the document
Facilitates the management of document classification within a defense project	

Table 6.1. *Using variables in requirements*

"Route energy" function

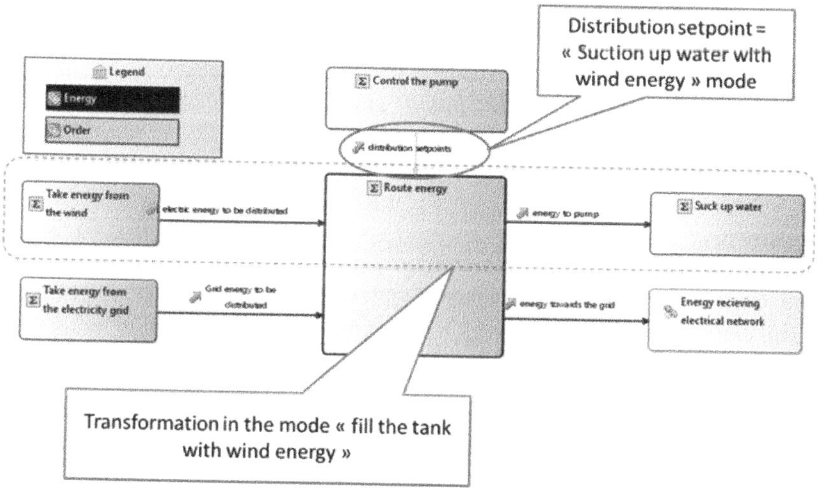

Figure 6.27. *Unifying thread example route energy function*

Requirement

The renewable energy water pump system shall route electrical energy to draw water from wind electrical energy if the distribution setpoint is "draw water with wind energy" in "fill tank and use wind energy" mode.

"Suck up water" function

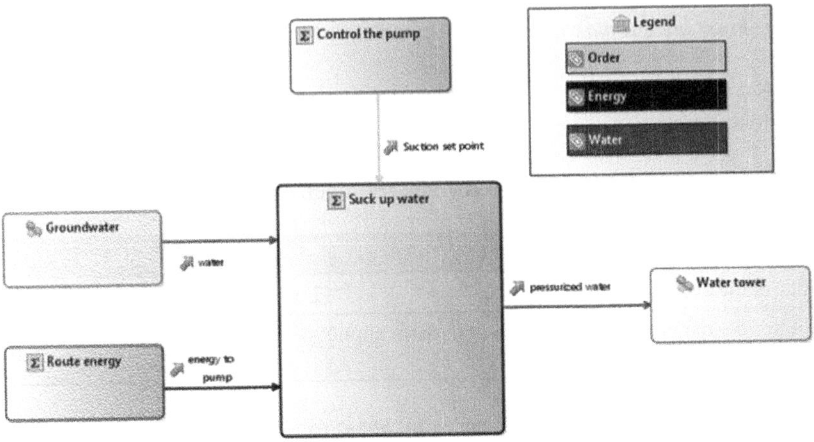

Figure 6.28. *Unifying thread example suck up water function*

Requirement

The renewable energy water pump system shall draw water from the groundwater with a maximum flow rate of 55 m^3/h in the "fill the tank" functional mode.

Requirement

The renewable energy water pump system shall draw water by converting electrical energy into pressure energy with a minimum efficiency of 50%.

Requirement

The "suck up water" function must be controlled by following the suction setpoint in the "suck up water" operating mode.

6.8. System specification

6.8.1. *The specification document*

The set of system requirements is organized in a specification document (see Figure 6.29).

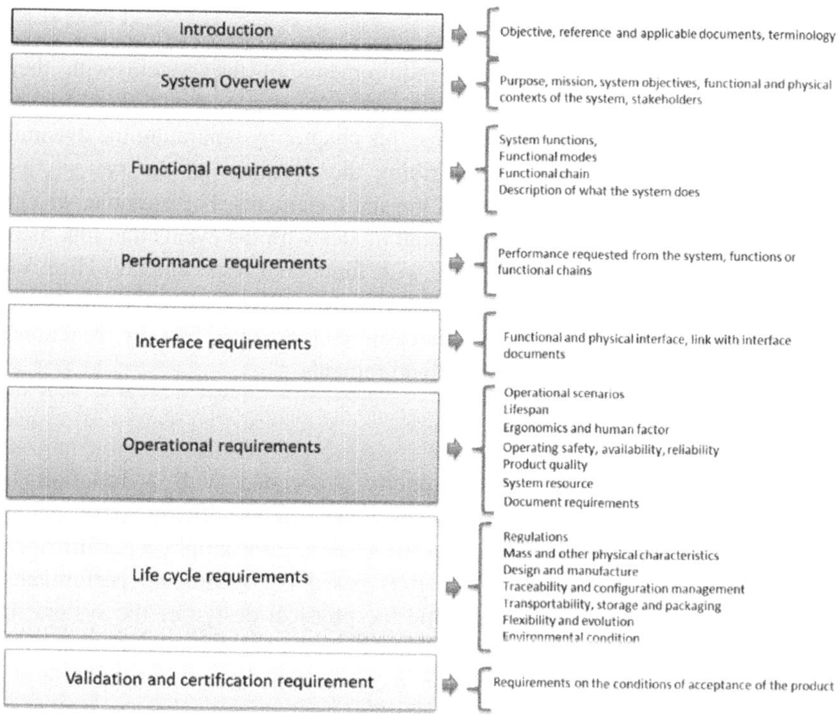

Figure 6.29. *Structure of a specification*

6.8.1.1. *Presentation of the system*

In a specification, the presentation of the system is also called the application domain and can be broken down into two parts. First, we can have a general description of the project with a presentation of the development context of the system.

Second, we have a general description of the system with:

– a presentation of the purpose of the missions and objectives of the system;

– a presentation of the functional context of the system with a description of the input and output flows of the system;

– a presentation of the physical or organic context with a description of the known interfaces of the external environment and the system;

– an overview of stakeholders and external systems.

6.8.1.2. *Functional requirements*

In this part, we put the functional requirements of the system with their associated performances. We need to prescribe what the system should do and when it should do it. For example, we can organize this chapter by separating the dynamic and static elements of the system. Prescribing the dynamics of the system first allows reference to it in the prescription of the static elements. For example, we can make a presentation of the system's functional modes with the events that link them together. We specify each operating mode with functional requirements. Then, we can list the functions of the system by presenting for each of them the inputs and outputs that characterize them. Each function can be specified by functional requirements with associated performance requirements.

6.8.1.3. *Performance requirements*

A performance requirement is a property associated with a **function** or **functional chain** or the **system**. It prescribes a quantifiable criterion of a quality attribute of a function, a functional chain or the system, for example, a performance, an accuracy or a duration (AFIS 2012). It is based on operational performance requirements. These requirements determine the physical design of the system. In this part, we put the performance requirements associated with the **system**. The performance requirements associated with the functions that can be put in the previous chapter.

6.8.1.4. *Interface requirements*

These requirements can be grouped by interface between the system and the external environment.

Each interface forms a subpart that can be divided into two by prescribing functional interfaces, i.e. flow exchanges, and physical interfaces, i.e. links between concrete elements (the system and a stakeholder or an external system). According to the EIA-632:1999 standard (ANSI 1999), an interface requirement specifies the conditions of interaction between elements.

EXAMPLE.– For the functional interfaces, we can specify the exchange scenarios of the flows as well as their characteristics.

EXAMPLE.– For physical interfaces, we can transform the interface constraints of the environment into concrete requirements on the system.

6.8.1.5. *Operational requirements*

Operational requirements include:

– operational scenarios: this part presents the mission profiles in the form of operational scenarios that will have to be covered by the system;

– lifetime: we have requirements on the lifetime of the system;

– ergonomics and human factors: we can identify the requirements of ergonomics and human factor consideration;

– dependability: we have the requirements of dependability with safety, reliability, availability and maintainability of the system;

– information security requirements;

– operational environment requirements: we have the requirements on the environmental conditions of the system;

– the means or resource requirements: here, we can identify the resource needs of the system (other than the interface requirements).

6.8.1.6. *Life cycle requirements*

The life cycle requirements are constraints translated into requirements for the system. We can have, for example:

– physical characteristics;

– the design and realization with, for example:

- studies and imposed solutions,
- materials,
- manufacturing,
- product identification;

– flexibility and evolution;

– transport, storage and packaging;

– commissioning;

– withdrawal from service;

– maintenance and documentation;

– production;

– regulations;

– product costs and lead times.

6.8.1.7. *Validation and certification requirements*

The list of requirements we have given is not exhaustive and depends on the system specified. However, we may also have validation and certification requirements.

6.8.2. *Application to the unifying thread example: system specification*

6.8.2.1. *Functional requirements*

For the functional requirements, we can organize this part according to the functional modes and functions of the system.

For example, for the system function "**drive the pump**":

Requirement of the "drive the pump" function

The system shall regulate the flow of water from the groundwater between 0 and D_max in ^m3/h to fill the tank.

6.8.2.2. *Performance requirements*

A performance requirement must have an impact on the sizing of the system. For example:

Performance requirement

The system must produce at least X KWh of electrical energy with a wind speed Vmax.

This requirement will size, for example, the wind generator of the system.

6.8.2.3. *Interface requirements*

For the specification of interfaces, we have the functional and physical interfaces. For example, a functional interface requirement will specify the behavior of the interface or the characteristics of a flow.

For example, a **functional interface requirement** on behavior:

The system shall measure the height of water in the tank every minute.

For example, a **functional interface requirement** on the characteristics of a flow:

The system shall measure the height of the water in the tank with an accuracy of plus or minus 1 cm.

Physical interface requirements specify the behavior or characteristics of a concrete link between the system and the external environment.

For example, a **physical interface requirement** on the characteristics of a link:

The system must be connected to the water inlet of the tank with the standard diameter of D_alim cm.

6.8.2.4. *Operational requirements*

In this part, we will find the system requirements which are the answers to the different operational constraints of the environment.

For example, a **maintainability requirement** could take the following form:

The system must be sustainable for at least 30 years. This requirement also applies to the entire supply chain.

For example, an **operational environment requirement** could be:

The system must have a minimum height of 3 meters between the ground and the tip of the wind turbine blade.

Finally, we can give an example of a **resource requirement:**

The system must operate on electricity supplied by the electrical grid.

6.8.2.5. *Life cycle requirements*

In this section, we will have the system requirements in response to the different constraints of the contributing systems throughout the life cycle.

For example, here is an example on a **physical feature requirement:**

The system must have a maximum height of X meters.

Or, we can give an example on a **design requirement on imposed solutions:**

The system must consist of a wind generator of at least X KWh.

6.9. Conclusion on requirements specification

The development of requirements can have two sources of input from the two families of stakeholders, namely the one interested in use and the other in realization. User stakeholders define the system's mission requirements and operational contexts. They can also express a set of constraints on the life cycle of the system. Stakeholders responsible for implementation express requirements for the system's life cycle and may identify opportunities related to user needs.

The system requirements are the expected properties of the system that can be classified into two categories, namely the functional requirements and the others that we simply call the non-functional requirements.

Direct translation of needs into requirements is possible, but would result in an incomplete and difficult to read specification. A functional analysis of the system must be done in parallel. This analysis is done on the basis of mission profiles, i.e. operational scenarios, and makes it possible to identify the functions, the functional modes and the interactions of matter, energy and information of the system.

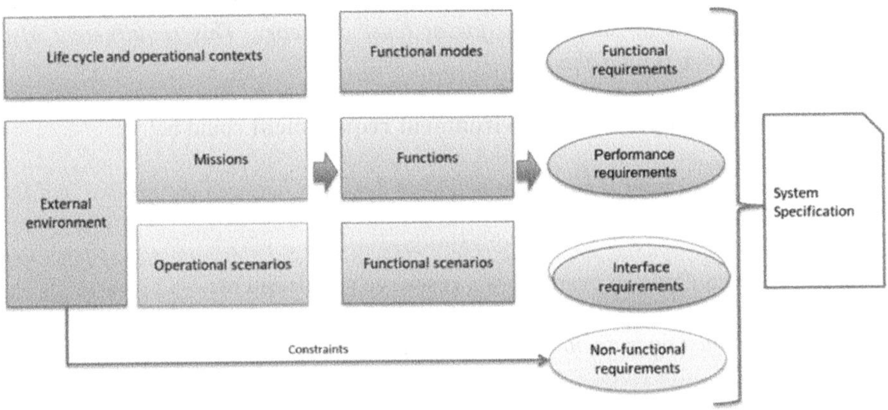

Figure 6.30. *Overview of the system specification process*

The system requirements are gathered in a specification which can be a document or a database. This specification is broadly organized as follows (see Figure 6.30):

– functional requirements with the functional modes, the functions of the system with the associated performance requirements;

– performance requirements with the overall system performance;

– interface requirements with MEI flow prescription and functional interface behavior as well as physical interface prescription with physical interface behavior;

– non-functional requirements with operational and life cycle requirements.

7

Requirements Validation

The requirements validation process takes place during the requirements specification process and makes the connection between requirements and needs.

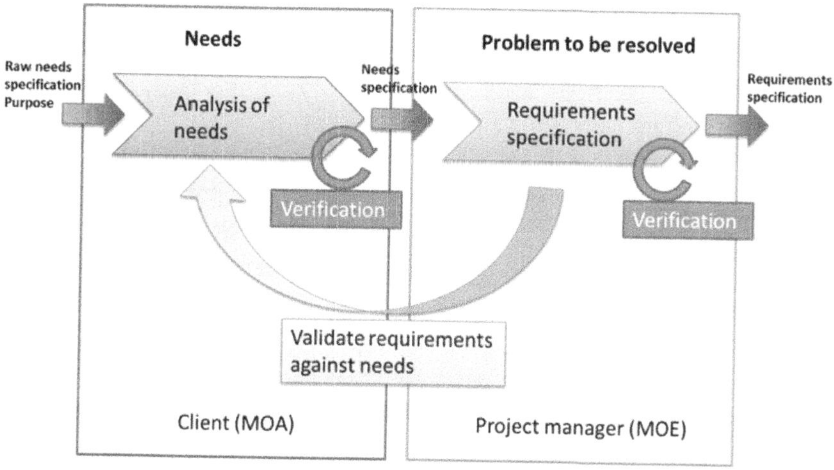

Figure 7.1. *V&V between needs analysis and requirements specification*

Figure 7.1 shows verification activities on the needs analysis and requirements specification process. **Verification** consists of checking that the requirements have been drafted according to the rules of the art. We can see the validation activity between the requirements specification process and the requirements analysis processes. **Validation** consists of looking at whether the requirements of the system

For a color version of all the figures in this chapter, see www.iste.co.uk/hygounenc/system.zip

Systems Engineering,
by Emmanuel HYGOUNENC. © ISTE Ltd 2023.

meet the needs and pose the problem correctly. Generally, when we talk about the requirements validation process, we include in the activity the verification of the requirements against the state of the art and the validation of the requirements against the needs.

7.1. General process

Figure 7.2 presents the requirements validation process inspired by the EIA-632:1999 standard (ANSI 1999). Each activity is presented in the form of a SIPOC with the elements required as input and the elements produced as output.

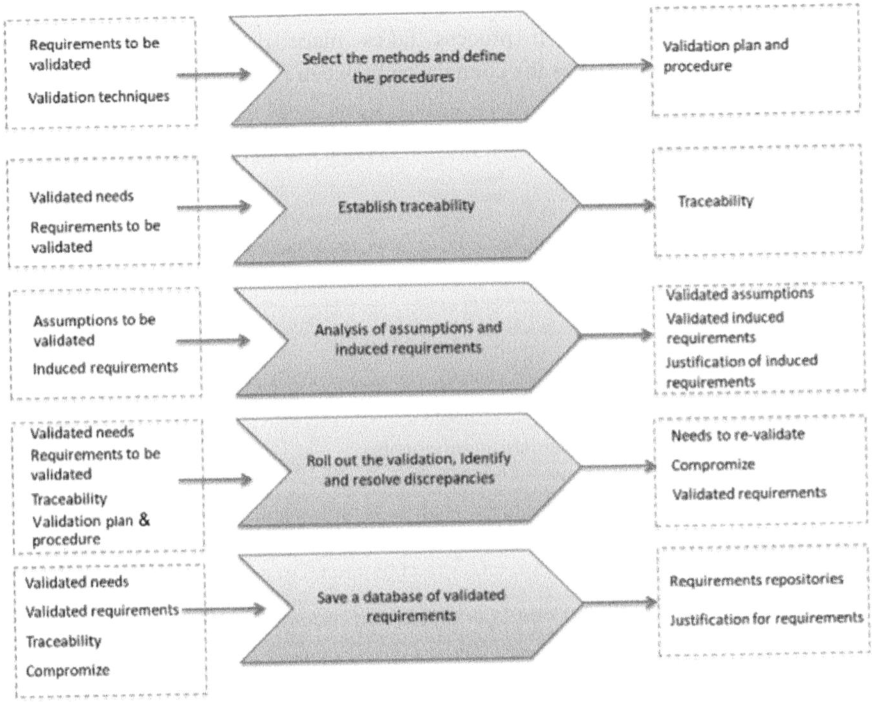

Figure 7.2. *Requirements validation process (ANSI 1999)*

The validation process is divided into several activities which are:

– the choice of validation methodologies;

– two-way traceability;

– analysis of assumptions and induced requirements;

– the validation process;

– registration of the validated repository.

7.2. Selecting methods and defining validation procedures

To validate a repository, we must define the validation strategy of the requirements vis-à-vis the needs. Several methods are available, and it is possible to apply several methods for the same requirement. The Aerospace Recommended Practice ARP4754A:2010 provides the following methods for requirements validation:

– vertical traceability;

– analysis;

– modeling;

– the test (this method integrates the simulation test);

– similarity;

– peer review.

7.3. Establishing requirements traceability

This activity must be done at the time of writing the requirements. The purpose of this activity is to verify that the traceability is correct.

7.3.1. Establishing traceability from needs to requirements

We need to establish downward traceability between needs and requirements. That is, each need must be able to be covered by at least one system requirement.

7.3.2. Establish traceability from requirements to needs

We need to establish upward traceability between requirements and needs. That is, each system requirement must cover at least one need. For the derived requirements having an impact on the system, the need comes from the designers and not from the customer. These requirements must be justified and accepted by the client. Depending on the domain, they must also be accepted by operational safety.

7.4. Analysis of assumptions and induced requirements

7.4.1. *Analysis of assumptions*

The assumptions formulated during the definition of the requirements must be managed in the same way as the requirements and in this context must be validated.

7.4.2. *Analysis of induced requirements*

During the design activity, derived requirements of this activity can be identified. An induced requirement is a requirement that cannot be traced back to a customer need. However, it must be traceable and justified vis-à-vis a need of the stakeholders interested in carrying out the project. The resulting requirements must be analyzed from a system and dependability point of view. If the requirement has an impact on the system or on the system environment, we must escalate it to the system repository. Clearly, this induced requirement becomes a property of the system. The resulting requirements must be analyzed from the point of view of operational safety in order to verify that they do not impact the environment. If this were the case, then it would have to be raised to the level of the supersystem, i.e. the customer so that they can make their own analysis of dependability.

EXAMPLE.– We are designing propeller propulsion. An electric motor was chosen to drive the propeller. This choice entails induced requirements on electromagnetic radiation and vibrations. If these requirements go back to the system level, the designers must warn their client so that they can verify that these new constraints have no impact on their aircraft.

7.5. Rolling out the validation

To roll out the validation of the system requirements, we can answer the following six questions, as proposed by the NASA System Engineering Handbook (NASA 2007).

– Are the requirements written correctly? That is, they include an applicant, a capacity, a performance and a context.

– Are the requirements technically correct? That is, the capacities requested are in accordance with the state of the art.

– Do the requirements meet the needs? That is, the capacities requested fulfill the missions and meet the constraints.

– Are the requirements feasible? That is, the capacities with the requested performance are realistic.

– Are the requirements verifiable on the integrated system? That is, the requested performances are testable on the integrated system.

– Are the requirements redundant or overspecified? That is, requirements are covered by others (i.e. less constraining) and that contexts of realization of the requirements are poorly expressed or not expressed at all.

– Are the requirements atomic? That is, they deal with only one subject.

The starting assumption of this process is that the needs have been verified and validated.

7.5.1. *Checking the requirements individually (correctness)*

The objective is to prepare the verification of the product against the system requirements. For this, we need to check that the requirements are SMART:

– **specific:** the requirement must specify a "to do" or "to be" capability;

– **measurable**: the requirement must have a measurable performance (criterion);

– **achievable**: the requirement must apply to a given context and must comply with the state of the art;

– **relevant**: the requirement must apply to the system or a function of the system. It can be applied to a component of the system in the case where there are particular constraints. However, it is better to stay at the system level and move the requirement down to the component specification level;

– **traceable**: the requirement must respond to an upstream request.

7.5.2. *Globally checking the requirements (consistency)*

The objective is to check the consistency of the requirements repository. To do this, we compare the requirements of the repository two by two in order to verify the following two points:

– uniqueness of requirements: each requirement of the repository must be unique;

– no contradiction: there must be no conflict between the requirements of the reference system

The quality requested from a requirements repository is summarized by the acronym “MACC” (AFIS 2012):

– **Maintainable**: the repository can be modified in a uniform way and it is possible to know and follow these evolutions.

– **Accessible**: the repository is accessible to all stakeholders entitled to know its content.

– **Complete**: the repository covers and meets all needs.

– **Consistent**: the requirements of the repository are compatible with each other and do not present any conflicts.

7.5.3. *Validating requirements against needs*

The objective is to validate the consistency between needs and requirements:

– validate the consistency of system inputs/outputs;

– validate the consistency of system performance against environment performance requirements;

– validate the consistency of the constraints imposed on the system.

7.6. Identifying and resolving gaps between needs and requirements

7.6.1. *Identifying gaps and conflicts*

In order to be able to resolve them by finding a compromise, we must identify the gaps and conflicts between needs and requirements. In particular, we need to identify gaps and conflicts in terms of:

– performance or performance allowance;

– compliance with constraints;

– consistency of external interfaces.

7.6.2. *Assessing conflicts*

We need to assess the conflicts between needs and system requirements. The conflicts identified in the requirements validation process are the entry point for this activity. We need to identify the alternatives in terms of functional and performance requirements. We need to evaluate and select a set of requirements and constraints with respect to their impacts on risk, cost, schedule and system performance.

7.6.3. *Study of trade-offs*

We must define the scope and conduct the study of the trade-offs between needs and requirements. For that, it is necessary to:

– identify and analyze the value of each alternative on the requirements;

– identify and quantify the risks of each alternative on the requirements;

– identify and analyze the cost of each alternative on the requirements.

7.6.4. *Identifying trade-offs and impacts*

Identify trade-offs and impacts between needs and requirements. Risk management options need to be identified. We need to select a set of requirements and constraints and document the following in the repository:

– the decision criteria;

– the justifications;

– the recommendations (requirements, assumptions and risks).

7.7. Saving a database of validated requirements

Following this evaluation and comparison to find compromises between needs and requirements, we can establish a validated requirements base. It must be ensured that this database contains the following elements:

– the system requirements;

– the risks associated with requirements;

– the assumptions;

– the justification for trade-offs;

the justification of the requirements

7.8. Conclusion on requirements validation

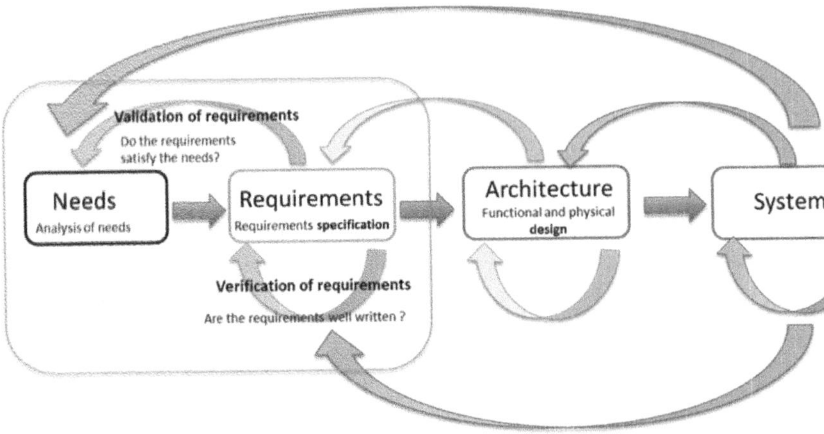

Figure 7.3. *Validation of requirements*

Figure 7.3 shows where the requirements validation activity takes place in the different systems engineering processes. Validation of requirements (see Figure 7.3) consists of looking at whether the problem has been well posed by answering the following question: Do the requirements satisfy the needs?

The answer to this question requires two prerequisites. The first is that the needs have been checked and validated by the project owner and the second is that the requirements have been checked by the project manager.

REMINDER.– the specification is achieved on the black box system. It brings together all the functional and physical properties of the integrated system. Functional properties respond to system missions from a static, discrete-event dynamic, and behavioral dynamic perspective. The physical properties (or non-functional in a more generic way) cover all the constraints imposed by the environment.

PART 4

System Design

8

Functional and Logical Design

8.1. Design and functional architecture

Functional design consists of studying an abstract solution of the system to meet the functions and missions that have been identified in the requirements definition. This design process consists of exploring one or more functional architectures from three points of view:

– static functional architecture, i.e. the study of the network of sub-functions;

– discrete dynamic functional architecture, i.e. the detailed study of the functional modes and associated events;

– behavioral functional architecture (continuous dynamics), i.e. the study of the dynamics of exchanges between sub-functions in a given operating mode.

8.1.1. *Static functional architecture*

A **static functional architecture** is the study of sub-functions and their interactions with each other and with the environment.

This static architecture is created according to a dual approach:

– an approach of decomposing functions into sub-functions on several levels to form trees;

– a reorganization approach that provides a structural view of internal functions and their interactions (see Figure 8.1).

For a color version of all the figures in this chapter, see www.iste.co.uk/hygounenc/system.zip.

Systems Engineering,
by Emmanuel HYGOUNENC. © ISTE Ltd 2023.

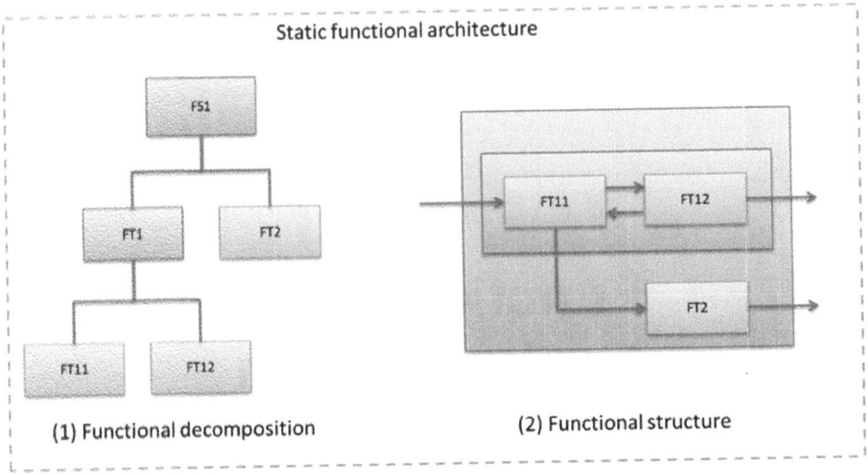

Figure 8.1. *Static functional architecture*

8.1.2. *Dynamic functional architecture*

A **dynamic functional architecture** is the study of the operating modes with their events as well as the activation of the sub-functions (Figure 8.2). The static functional architecture represents all possible functional interactions. In the operating modes, the functions are activated or deactivated according to the desired behavior.

For example, in Figure 8.2, in the requirements specification part, we have two functional modes, with the "F1" function activated in "mode 2". In the functional design, the static architecture breaks down the "F1" function into three sub-functions. These sub-functions form two functional chains "F11-F12" and "F11-F13" which are respectively activated in the sub-modes "M21" and "M22".

Each sub-function thus activated transforms incoming flows into outgoing flows. For each sub-function, the need to be controlled or regulated must be analyzed through the behavioral point of view. Figure 8.3 shows a transformation type function S = F(E) as in Figure 6.9, but adding the optional flows needed for regulation. Indeed, to be able to regulate a function in closed loop, it is necessary:

– to measure these effects through the outgoing control flows, in order to calculate the control instructions;

– to transmit the control instructions through the incoming control flows, in order to act on the incoming and outgoing flows. Incoming control flows also enable functions.

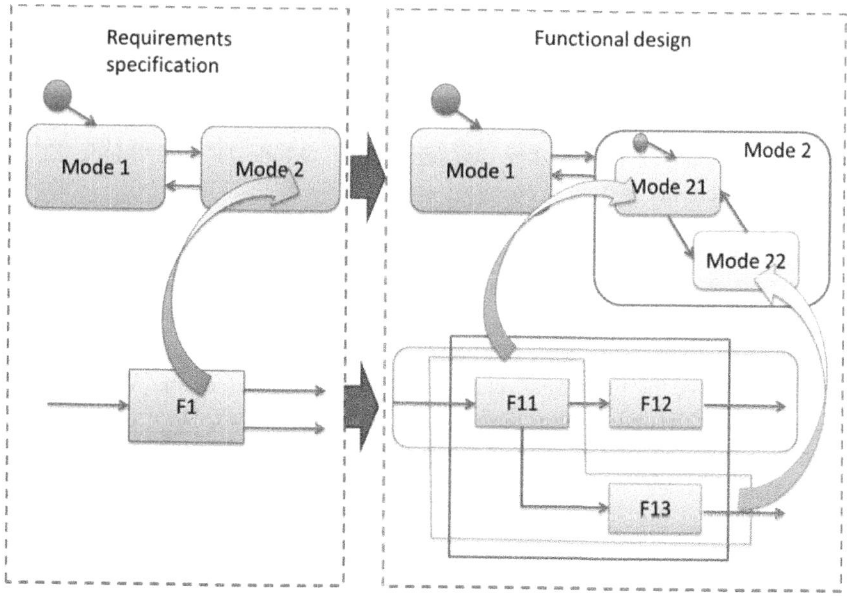

Figure 8.2. *Refinement of functional modes*

It is also possible to control a function in open loop by sending the setpoints through the incoming control flows. This also enables the function. In the case of an open loop, either there are no outgoing control flows or they are not used for regulation.

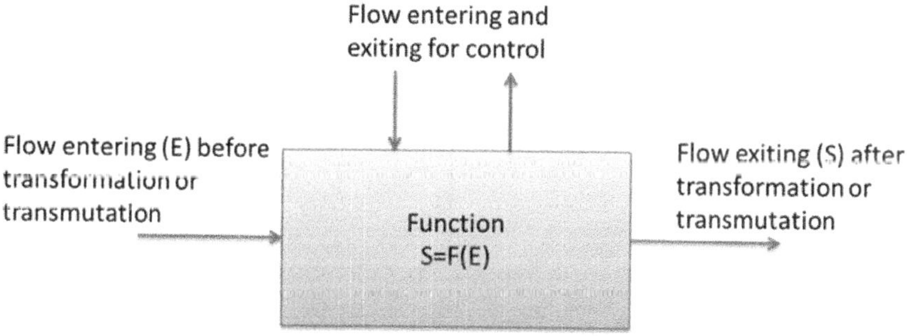

Figure 8.3. *Function with control*

8.1.3. *Behavioral functional architecture*

A **behavioral functional architecture** or **continuous dynamic** architecture is the integration of static and dynamic discrete event views to reconstruct system functions and missions in operational contexts.

In Figure 8.4, we have a static architecture of three internal functions of the system as well as a dynamic architecture with two functional modes of which "Mode 2" has two functional sub-modes. The two sub-modes, "Mode 21" and "Mode 22", are activated respectively by two events evt1 and evt2.

At the bottom of Figure 8.4, we have the behavioral view which is divided in two. On the left side, we have the execution dynamics of the chain of sub-functions. The control flows allow the activation of the different functions. We can see that the functions "F12" and "F13" are conditioned by the events "evt1" and "evt2". On the right side, we have the functional scenario of the three sub-functions. We can see the alternatives conditioned by the events "evt1" and "evt2" which correspond to the respective activation of the sub-functions "F12" and "F13", corresponding to the functional sub-modes "Mode 21" and "Mode 22", respectively.

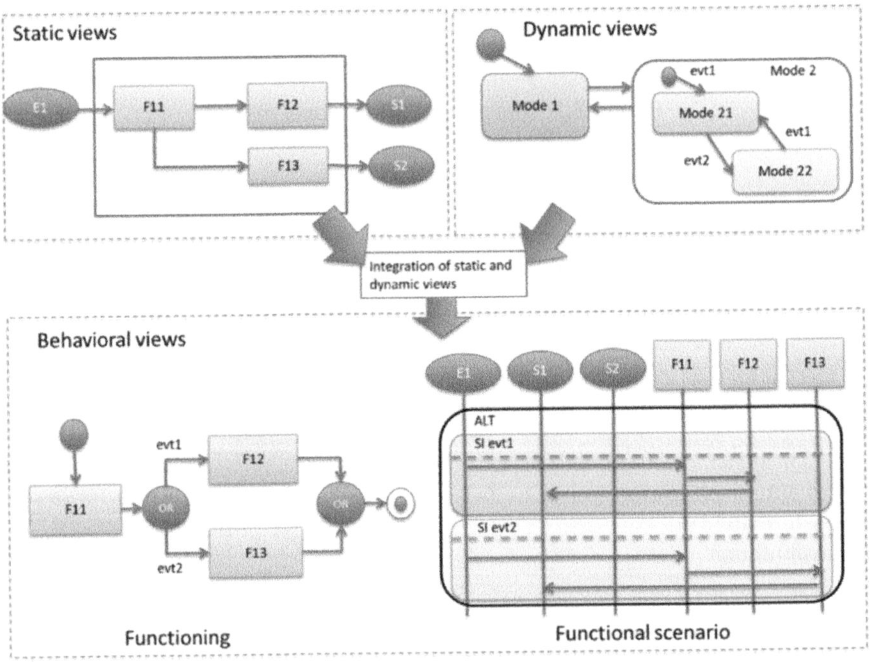

Figure 8.4. *Behavioral view (continuous dynamics)*

8.2. Identifying the sub-functions of the system and their interactions

8.2.1. *Purpose of identifying sub-functions*

The objective of this activity is to decompose system functions into sub-functions until they can be allocated to system components. The **sub-functions** are the **internal** functions of the system that we can also call the **technical functions**. As well as the system functions:

– a technical function transforms incoming flows into outgoing flows;

– it is activated in operating modes;

– it can be controlled.

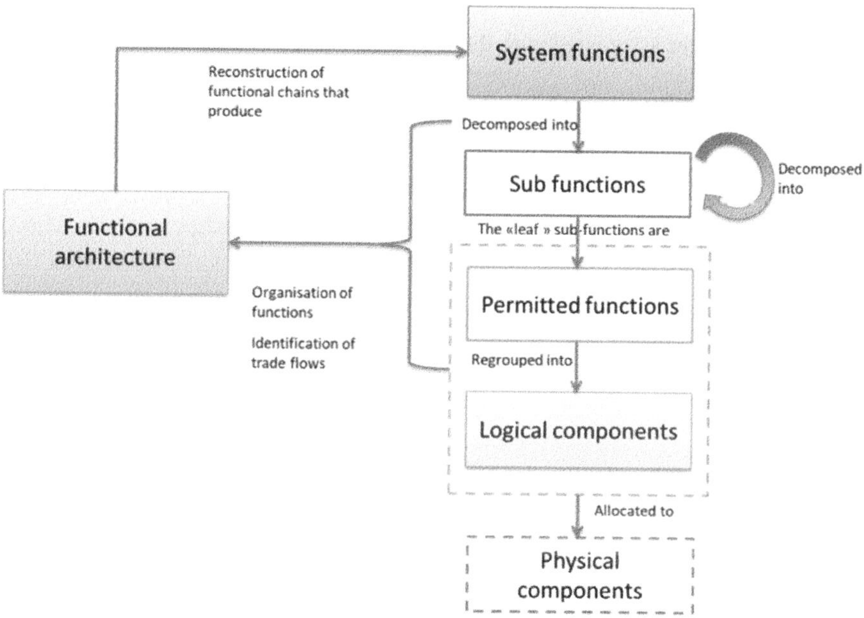

Figure 8.5. *Functional decomposition*

This activity of identifying the technical functions is a solution to the problem posed by the system functions and the associated requirements. We have two outputs from this analysis, namely a functional tree (see Figure 8.5) and a structural organization of the leaf sub-functions with their internal and external interactions (the MEI, material, energy and information flows).

8.2.2. *Functional chain analysis approach*

There are several functional analysis methods such as the two classical and complementary approaches which are FAST and SADT (Tassinari 2006; Audry 2010). We will present another approach based on the analysis of functional chains using a framework. This approach has the advantage of making the link with the analysis of the system functions of the requirements specification process and allowing the functional decomposition in a systematic way.

Moreover, it allows both to make the functional decomposition while building the structural view. As an input to this activity, we need (see Figure 8.6):

– system functions with their incoming and outgoing flows and associated requirements;

– the environment associated with the functions;

– functional chains.

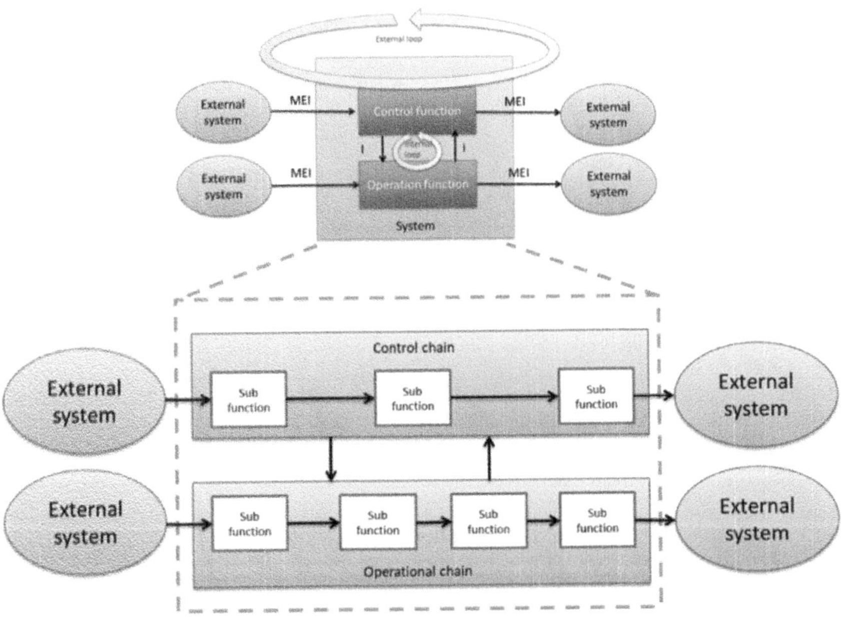

Figure 8.6. *Decomposition of functions and functional chains*

The functions of the system in charge of the control are made up of functional chains that process the information. The operation functions are made up of functional chains that transform energy. They can also transport energy or store it for later release.

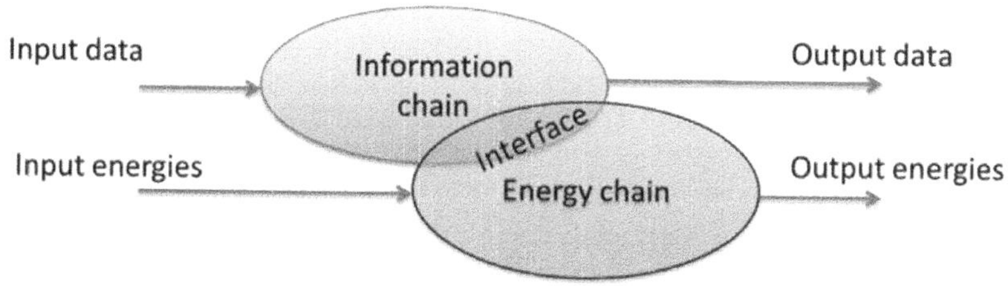

Figure 8.7. *The functional chains*

The two functional chains, information and energy, are in interface and exchange information for the regulation (see Figure 8.7). This view is an abstraction of the OID model of Le Moigne (see Figure 1.31).

The information chain can be broken down into acquisition, processing and communication functions, as shown in Figure 8.8. The acquisition and communication functions are interface functions of the functional chain.

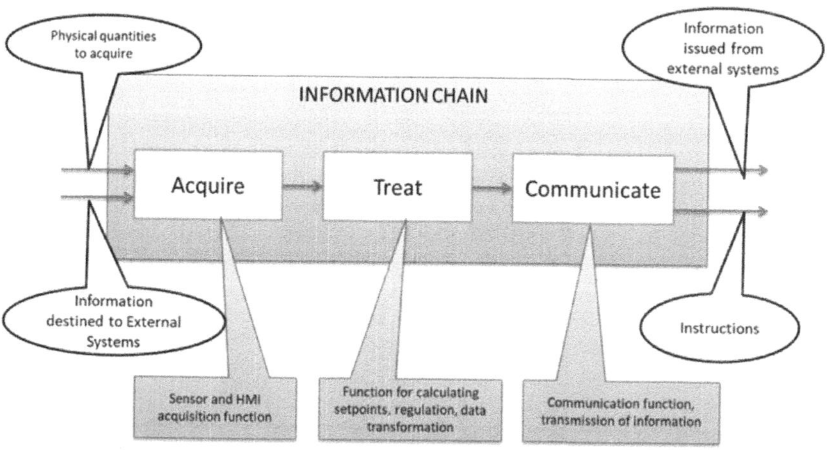

Figure 8.8. *Information chain*

The energy chain can be broken down into supply, distribution, conversion and transmission functions (see Figure 8.9). The supply and transmission functions are the interface functions of the chain. The distribution functions interface with the information chain and take as input the instructions calculated by the information chain.

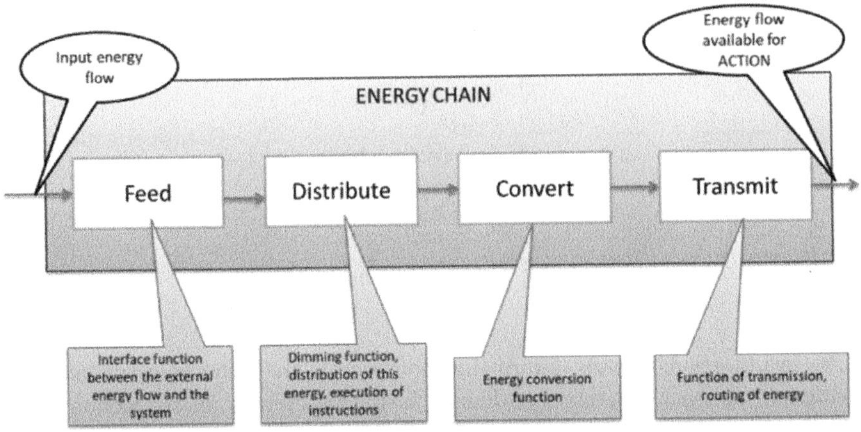

Figure 8.9. *Energy chain*

Figure 8.10 shows the integration of the two functional chains together. The processing function sends the setpoints via the communication function to a distribution function that regulates the energy input of the conversion function. The loopback can be done by the acquisition function on one or more functions of the energy chain.

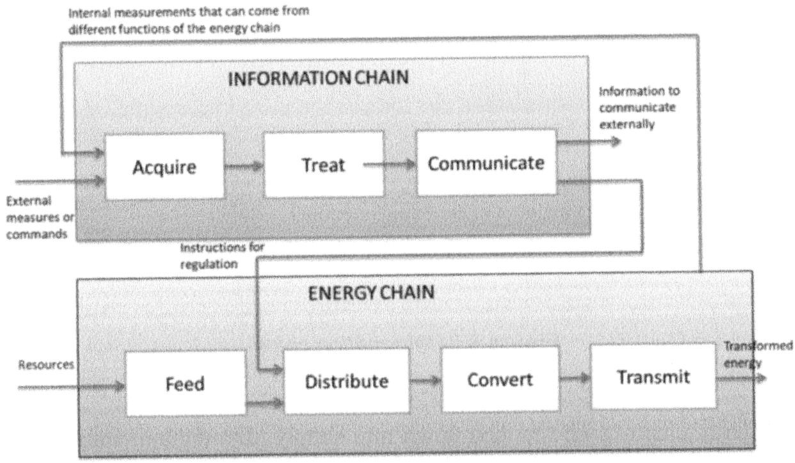

Figure 8.10. *Two interacting chains*

The MEI–TEF table in Figure 8.11 shows the projection of the two functional chains onto the elementary "transformations" and flow types. We can note that the functions of supply and acquisition will, according to the context, be able either to bring as value a change of time or a change of space.

	M material	E energy	I information
T time	Feed (Store)	Feed (Accumulate)	Acquire (Save)
E space	Feed, Transmit (Carry)	Feed, Transmit	Acquire, Communicate
F shape	Convert (Transform), Distribute	Convert, Distribute	Treat

Figure 8.11. *MEI–TEF matrix applied to the information/energy chains*

This approach of identifying and decomposing the functional chains of information and energy can be used as a checklist or for guided brainstorming. Indeed, the idea is not to follow this functional analysis framework to the letter, but rather to use it as a tool to initiate or guide the functional decomposition.

8.2.3. *Implementation of the unifying thread example: identification of sub-functions*

For each function of our system, we will identify the sub-functions to realize them. This decomposition tries to implement the methods that have been presented. As a reminder, the use of methods on functional chains is used as a checklist and not in a strict manner.

8.2.3.1. *Decomposition of the "control the pump" function*

In Figure 8.12, we have represented the decomposition in sub-functions of the function "control the pump". We have identified two acquisition functions which are "measure the water level" and "measure the wind speed" as well as a processing and communication function which is "regulate the filling of the water tower".

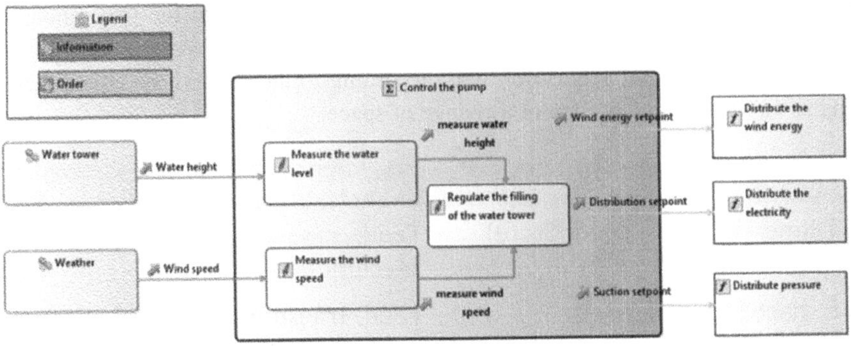

Figure 8.12. *Decomposition of the "control the pump" function*

8.2.3.2. *Decomposition of the function "take energy from the wind"*

In Figure 8.13, we have the decomposition into sub-functions of the function "take energy from the wind". We have identified a supply function that is "capture wind energy", a distribution function that interfaces with the regulation which is "distribute wind energy", a conversion function that is "convert wind energy into electrical energy" and a transmission function that is "transmit wind electricity".

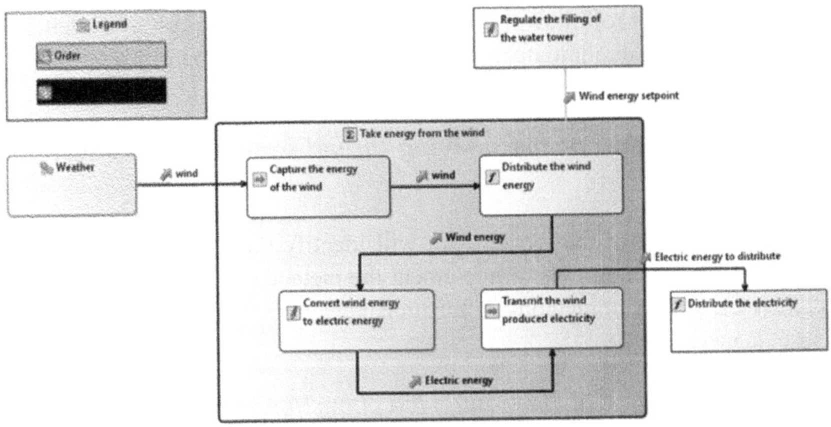

Figure 8.13. *Decomposition of the function "take energy from the wind"*

8.2.3.3. *Decomposition of the function "take energy from the electrical network"*

In Figure 8.14, we have the decomposition into sub-functions of the function "take energy from the electrical network". We have identified a transmission function that is "transmitting electricity from the grid".

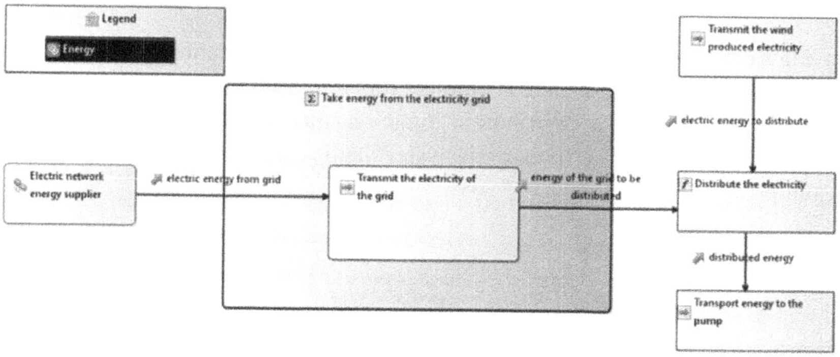

Figure 8.14. *Decomposition of the function "take energy from the electrical network"*

8.2.3.4. *Decomposition of the "route energy" function*

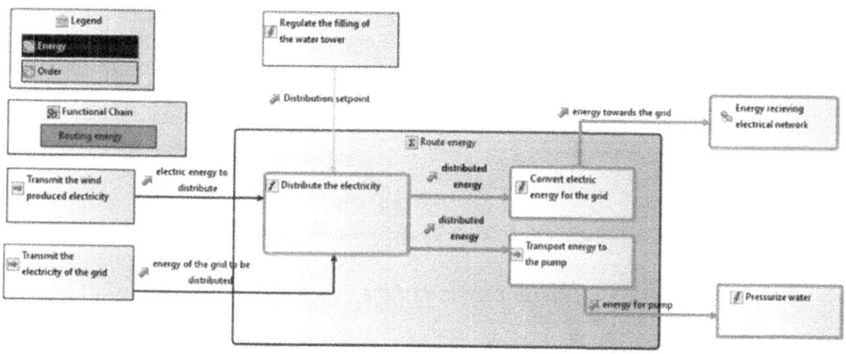

Figure 8.15. *Decomposition of the "route energy" function*

In Figure 8.15, we have the decomposition into sub-functions of the function "route energy". We have identified a distribution function that is "distribute electricity" as well as a conversion and transmission function that is "convert electrical energy for the grid" and another transmission function that is "transmit energy to the pump". These three sub-functions form the functional chain of the routing of the electric energy, which aims to take in entry the various sources of electric energy and, according to the mode of functioning, to direct them either towards the function of pressurization of the water of the water table or towards the outside of the system, namely, the electric network. This functional chain is driven by the system's control function, which is "control the filling of the water tower".

8.2.3.5. *Decomposition of the "suck water" function*

In Figure 8.16, we have the decomposition into sub-functions of the function "suck water". We have identified a complete functional chain with a supply function that is "supply water", a distribution function that is "distribute pressure", a conversion function that is "pressurize water" and a transmission function that is "transmit pressurized water".

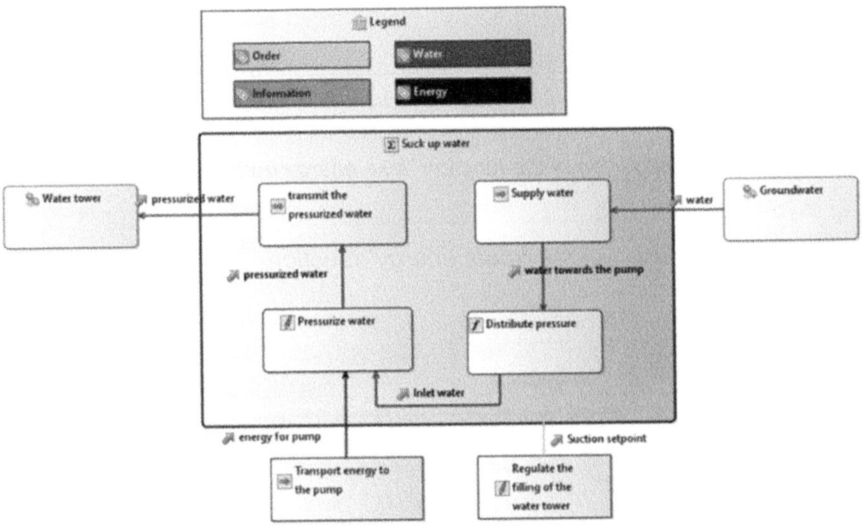

Figure 8.16. *Decomposition of the "suck water" function*

8.2.4. *The outputs of the static architecture*

8.2.4.1. *The functional tree*

The functional decomposition can be in the form of one or more trees depending on the complexity of the system. We have several possible representations. We can start with a system mission and show how it is covered by the system functions and sub-functions. We can start from the system itself and show what its functions and sub-functions are. This representation can give rise to a tree structure if we have an exclusive decomposition of functions into sub-functions. If this is not the case, it is better to make several trees. We can also start from the system with the decomposition of its sub-functions, i.e. its internal functions and then go back to the system functions by grouping and abstracting the sub-functions to form a tree structure. The functional tree is called the FBS, "Functional Breakdown Structure". This is an important output for the organization of the project. Indeed, this tree structure allows the allocation of project tasks and resources.

8.2.4.2. *Application to the unifying thread example: the FBS*

In Figure 8.17, we can see a tree representation of our unifying thread example. The tree structure is presented in the form of Russian dolls to facilitate reading.

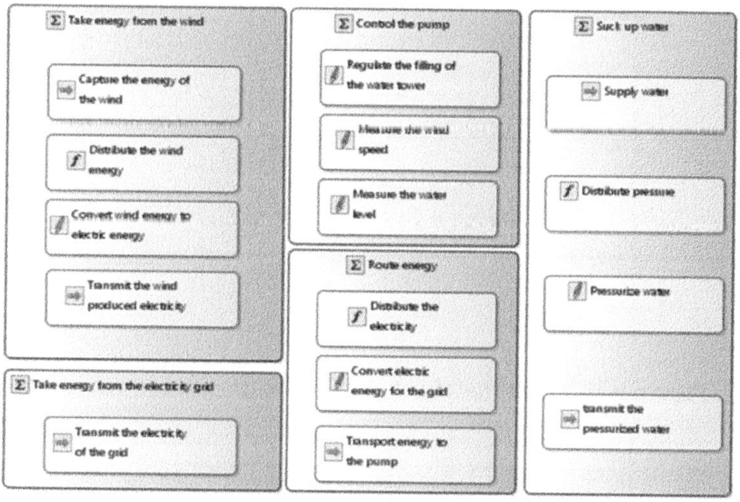

Figure 8.17. *Functional decomposition of the unifying thread example*

Regarding our unifying thread example, we have two levels of functional decomposition. The first level represents the functions of the system that are specified by the associated functional and performance requirements. On the second level, we have the sub-functions resulting from the design work, which are the internal or technical functions of the system. The starting point is our "renewable energy water pump" system.

8.2.4.3. *The structural view*

The objective of this view is to present the functional structure of the system on one or more diagrams. We can have on these diagrams the sub-functions of the system with their interactions in terms of MEI flows, i.e. material, energy and information. To maintain the big picture, we can add external systems or stakeholders.

8.2.4.4. *Application to wire: functional structure*

Our "renewable energy water pump" example has been deliberately simplified so that all its sub-functions can be presented in a single diagram. The diagram showing the functional structure of the water pump system with renewable energy without control can be seen in Figure 8.18.

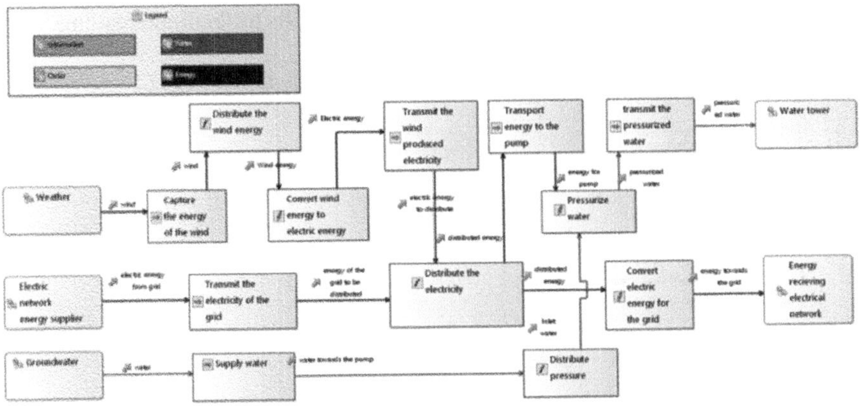

Figure 8.18. *Functional structure of the unifying thread example*

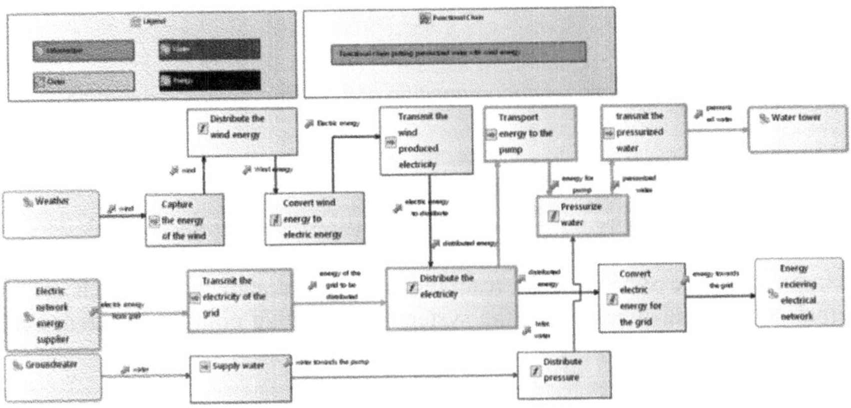

Figure 8.19. *Filling the water tower using wind energy*

This diagram shows only the sub-functions that contribute to the realization of the operation functions. In order to maintain readability, we have not represented the steering sub-functions.

In Figure 8.19, we show the architecture of the operation sub-functions, highlighting the functional chain that allows the water tower to be filled by pressurizing the water using the wind energy.

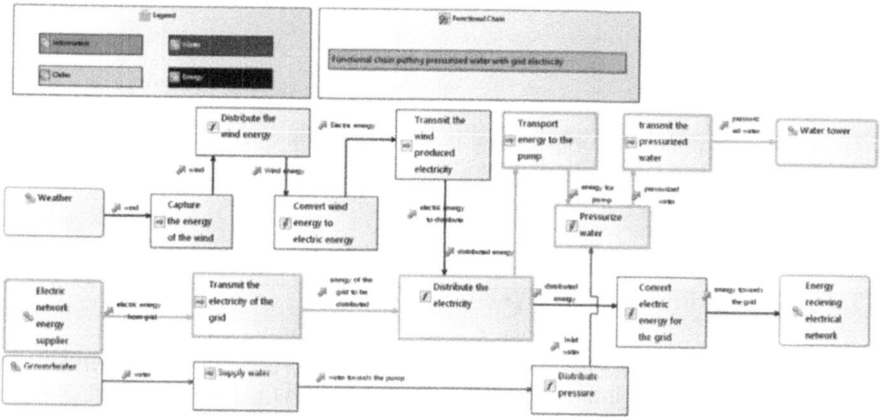

Figure 8.20. *Filling the water tower using energy from the electrical network*

In Figure 8.20, we show the architecture of the operation sub-functions, highlighting the functional chain that allows the water tower to be filled by pressurizing the water using energy from the electrical network.

8.2.5. Rules on the quality of a functional decomposition

8.2.5.1. *Stopping criteria*

We decompose the functions of the system into sub-functions until they can be allocated to components to build the solution. According to Meinadier (1998) and Micouin (2014), we can have the following four **stopping criteria**:

– the sub-function can be performed by a single trade or engineering;

– the sub-function has no more internal states, its behavior is defined by its inputs and outputs independently of external events. This is the definition of an atomic function;

– the sub-function can be allocated to a component developed in-house by the company. It is the decision of the project to do the realization in-house;

– the sub-function can be allocated to a component that can be purchased off the shelf. It is a project decision to "make", that is, to sub-contract the realization outside the company.

8.2.5.2. *Quality criteria*

A good functional decomposition is the result of a compromise where the designer will try to have, for example, a balanced allocation of the performances of the system functions on the sub-functions (Meinadier 1998).

8.3. Functional interface analysis

During the functional decomposition, we brought out new interactions between the sub-functions. These interactions are composed of all the flows of matter, energy and information transformed through the network of sub-functions. The objective of this activity is to characterize these new flows and to consolidate the flows already identified in the system specification. As an input to this activity, we have the functional decomposition with the flows exchanged between the sub-functions. At the end of this activity, we have the list of characterized flows that allow us to specify the system interfaces.

8.3.1. *The coupling matrices*

To analyze and visualize the system from the perspective of these functional interfaces, we can use the coupling matrix, also called the N2 matrix or DSM, Design Structure Matrix (Eppinger and Browning 2012).

This mathematical modeling tool represents in matrix form the couplings of a network of functions. The construction of the functional DSM is relatively simple (see Figure 8.21), which makes it an accessible and powerful tool. All functions of the modeled decomposition level, for example, the leaf functions of the decomposition, are plotted on the rows and columns. Then, each row corresponds to the outputs of the considered element and each column to the input of the corresponding element. Each intersection between a row and a column is a potential interaction. The functional network analysis allows the matrix to be completed. We can build several types of functional DSMs, namely:

– Boolean DSMs, which materialize the presence or absence of a functional exchange;

– DSMs that count the number of interactions between functions;

– DSMs with four cells: three that separate and account for the flows of matter, energy and information. The fourth cell is reserved for spatial interactions.

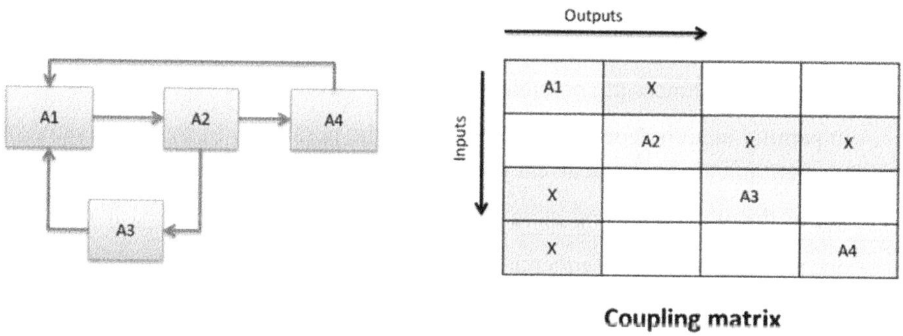

Figure 8.21. *Functional coupling matrix*

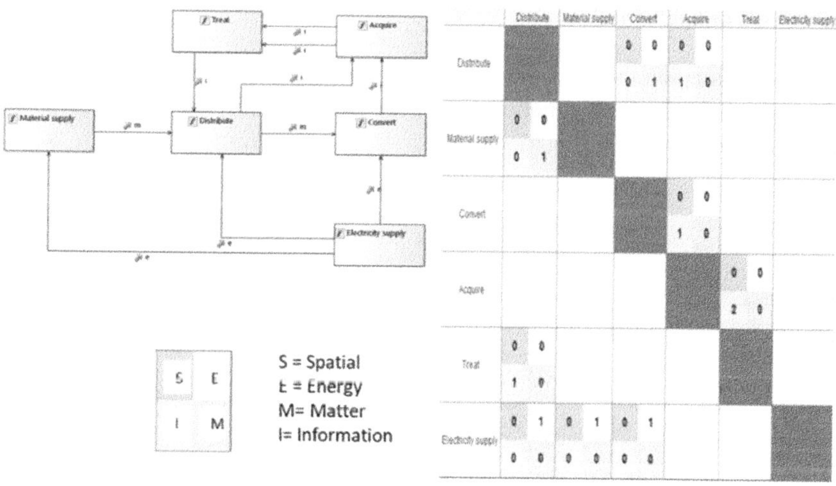

Figure 8.22. *Four-cell DSM*

Figure 8.22 shows on the left a functional architecture with functions that exchange flows of matter, energy and information. On the right side, we can see the representation of this architecture as a four-cell coupling matrix. In this example, the cell that represents the spatial relationships is not used.

For functional networks, the analysis of couplings identifies the possible choices in terms of dynamic behavior. For example, in Figure 8.23, we have the three cases that we can meet in a functional network.

The case of decoupled functions

In the example, the two functions A1 and A2 have no interaction. At the dynamic level, all choices are possible. They could be activated:

– in parallel in an independent way, i.e. with the possibility of being activated in distinct operating modes (concurrent modes);

– one or the other at the end of an alternative;

– one after the other in a sequential manner.

The case of functions in sequence

In the example, function A1 exchanges at least one flow with function A2. Both functions could be activated sequentially.

The case of coupled functions

In the example, the two functions are strongly coupled by having mutual exchanges of flows. Both functions could be activated in parallel in a dependent way, i.e. in the same operating mode.

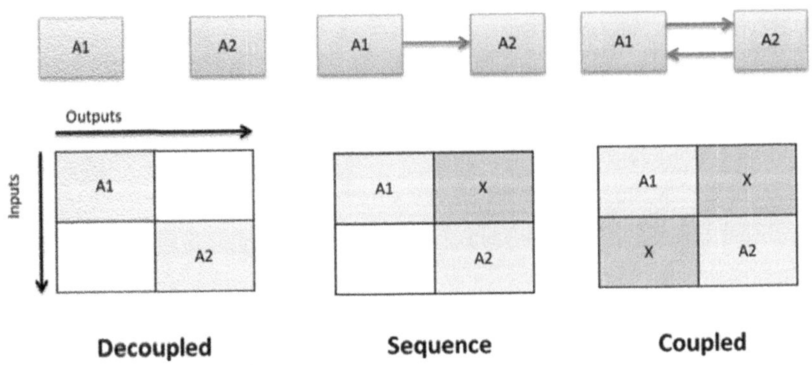

Figure 8.23. *Coupling analysis*

8.3.2. *Characterizing functional interfaces*

The set of functional flows must be characterized. This characterization will enable the definition of documents and interface requirements. Functional interfaces can be characterized with the following key elements, namely (Krob 2017):

– the nature of the flow, i.e. matter, energy or information;

– the purpose of the interface, i.e. which flows should be exchanged;

– the operation of the interface, i.e. how the flows are exchanged;

– the operating modes of the interface, i.e. when the flows are exchanged and when they are not;

– functional constraints, i.e. the performance associated with the interface.

Data that passes through the functional interface can be characterized in a table instead of textual interface requirements. The table in Figure 8.24 provides an example of a parameter list for specifying data for an interface (Baeckelandt 1992).

Name	Description	Source	Target	Unit of measure	Ladder	Precision	Resolution
Speed	Translation speed standard	Navigation software	Control software	m/s	[0 50]	+/- 0.1	0.001
Altitude	Height along the Z axis of the absolute reference	Navigation software	Guidance software	m	[0 300]	+/- 0.5	0.01

Figure 8.24. *Data table (Baeckelandt 1992)*

In this example, the knowledge of the source and destination comes from the physical architecture of the system when the sub-functions, which produce and consume the data, have been allocated to physical components.

8.4. Consolidation of functional modes

8.4.1. *Objective of consolidating functional modes*

The objective is to analyze the functional modes of the system and, depending on the situation, to break down the functional modes into sub-modes. This refinement is done from the analysis of the behavior of the system functions which is obtained by the activation of its sub-functions (Figure 8.25).

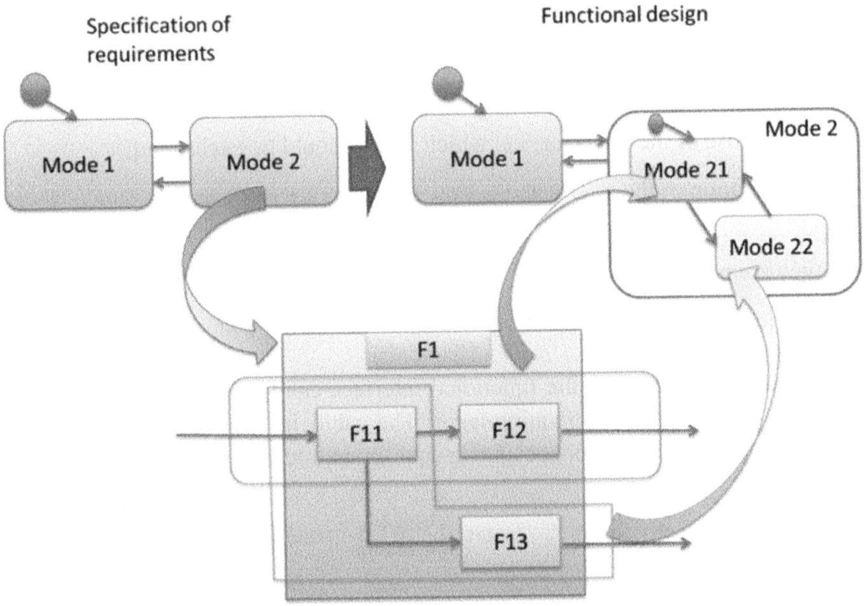

Figure 8.25. *Analyze modes and functions*

8.4.2. *What to do to identify the functional sub-modes*

As a reminder, during the specification of the requirements, the identification of the functional modes was done by taking into account the operational contexts and the missions of the system. The analysis of the operational contexts and missions allowed us to identify the functional modes with the associated system functions. We also recall that a functional mode translates a functional behavior of the system over a certain period of time. In this activity, the functional modes with the sub-functions of the system functions are taken over. When decomposing the functions of the system, we can have several functions that are represented by functional chains.

The operation of a function is the algorithmic study of the function through its sub-functions. This analysis, which is part of the functional behavioral architecture, also allows us to work on and refine the functional decomposition. There may be ways of functioning that change according to the context. Indeed, for the execution of a function of the system according to the internal and/or external events, there can exist several sequences of sub-functions. If these sequences give functionings that are exclusive, then the functional modes can be refined into sub-modes. To each sub-mode we will allocate the sub-functions activated for the operation of the system function, corresponding to internal and/or external events (Figure 8.26).

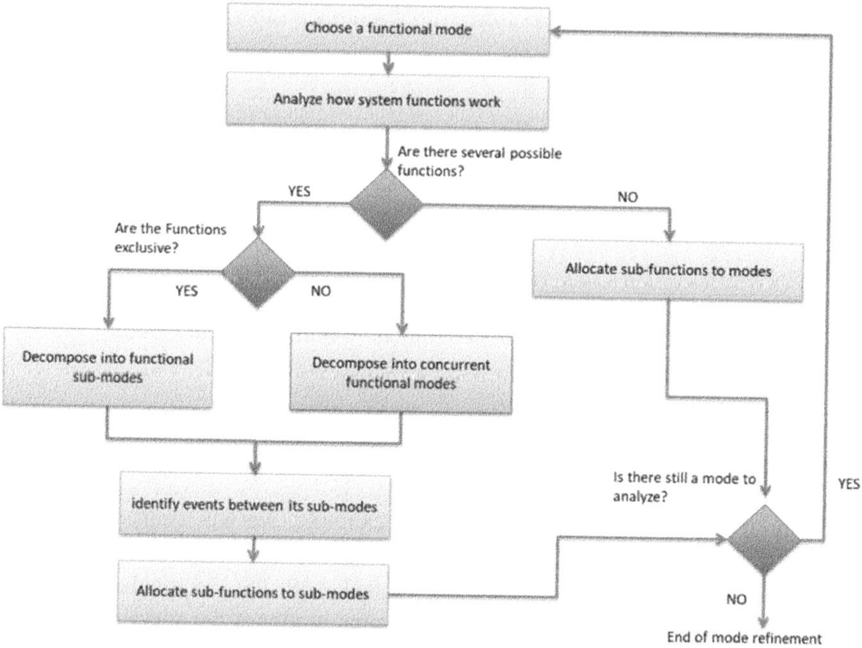

Figure 8.26. *Functional mode analysis process*

The process in Figure 8.26 identifies the functional sub-modes that will run sequentially or iteratively, i.e. the operations are activated in an exclusive way. This process also allows the identification of concurrent sub-modes, i.e. sub-modes in which operations are activated in parallel and are independent, i.e. they do not need to be synchronized.

There can be systems with a combination of both. That is, sequential or iterative sub-modes that are concurrent with sequential and iterative sub-modes (see

Figure 8.27). This decomposition into functional sub-modes controls the complexity of the system by facilitating its understanding. This decomposition into functional sub-modes may have already been identified during the specification of the system requirements.

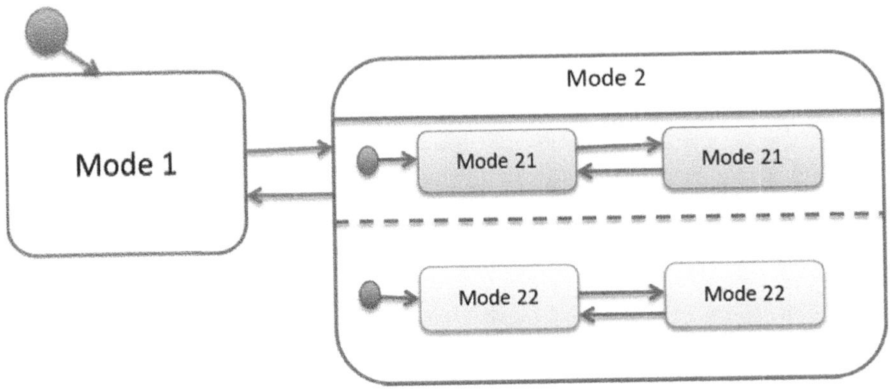

Figure 8.27. *Example of a state machine with functional modes*

8.4.3. *Application to the unifying thread example: resumption of the functional modes*

For the application to our unifying thread example, we will study the "fill the tank" functional mode. This functional mode is already refined into two sub-modes "fill the tank with wind energy" and "fill the tank with the electrical grid". Since processes are iterative, the refinement of functional modes at the specification level depends on the operational contexts and the system.

8.4.3.1. *Analysis of the "fill the tank with wind energy" mode*

In Figure 8.28, at the top we have the functional mode status machine with the the functional sub-mode "use wind energy". At the bottom of the figure, we have the block diagram with the sub-functions of the system operation chains. We have highlighted in blue the functional chain that pressurizes the groundwater with wind energy.

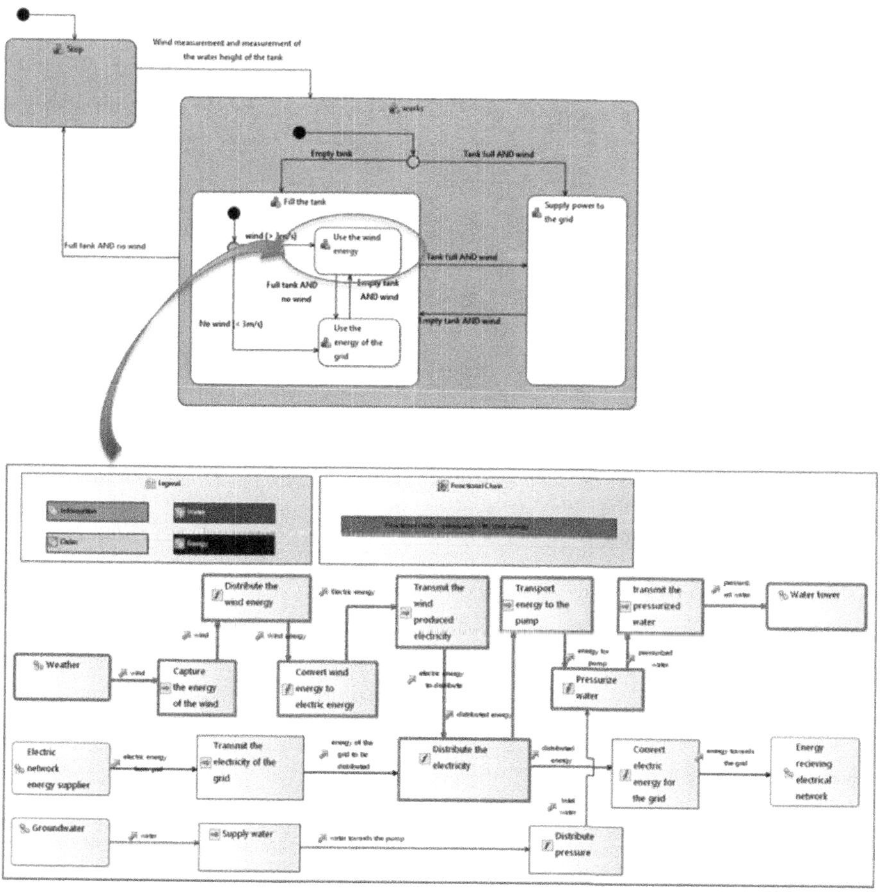

Figure 8.28. *Resumption of the "fill the tank with wind energy" mode*

8.4.3.2. *Analysis of the "fill the tank with energy from the network" functional mode*

In Figure 8.29, we have at the top the state machine of the functional modes with the functional sub-mode "use the energy of the network". At the bottom, we have the block diagram of the sub-functions for the system operation chains. We have highlighted in blue the functional chain that pressurizes the groundwater using the energy of the electrical network.

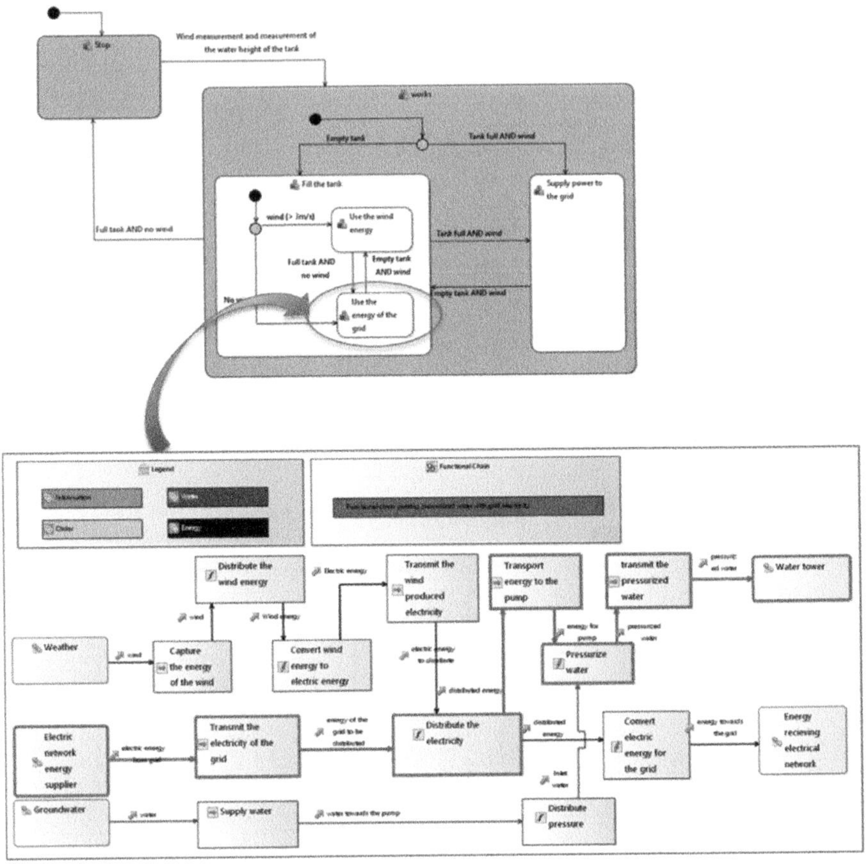

Figure 8.29. *Resumption of the "fill the tank with energy from the network" mode*

8.5. Resuming system functioning

8.5.1. *Objective of the system recovery*

In this analysis, there are two types of functioning to consider:

– the functioning of the functions of the system: that is, the algorithmic study of the function through its sub-functions;

– the functioning of the system: that is, the study of the operational scenarios, which are profiles of missions, with the sub-functions of the system.

REMINDER.– A mission profile is a mission in a given operational context.

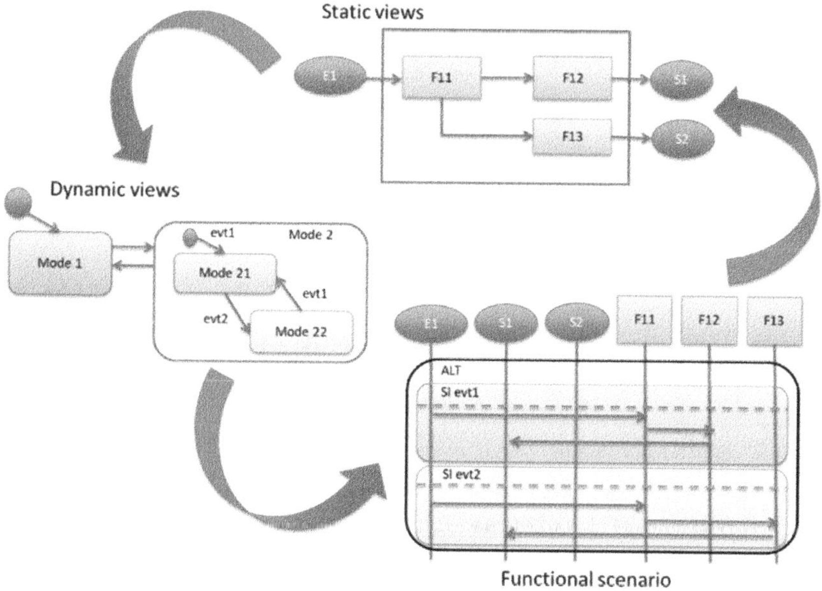

Figure 8.30. *Points of view are synchronized*

This behavioral analysis is the integration of the previous static and dynamic analyses. The system architecture proposes a study of the complex system by points of view. All of these points of view represent the system. These points of view are linked and synchronized with each other (see Figure 8.30). In Figure 8.30, we can note that the behavioral view is indeed the integration of the static and dynamic views. In fact, in the functional scenario, we have an alternative operation with two conditions each triggered by a respective event:

– on event "evt1", we find the behavior in "Mode 21" with the functions "F11" and "F12";

– on event "evt2", we find the behavior in "Mode 22" with the functions "F11" and "F13".

Furthermore, using a sequence representation, we can correctly identify:

– the sub-functions and the environment: a global systemic view;

– exchanged flows specific to the static view;

– the events characterizing the passage in the functional modes specific to the discrete dynamic view.

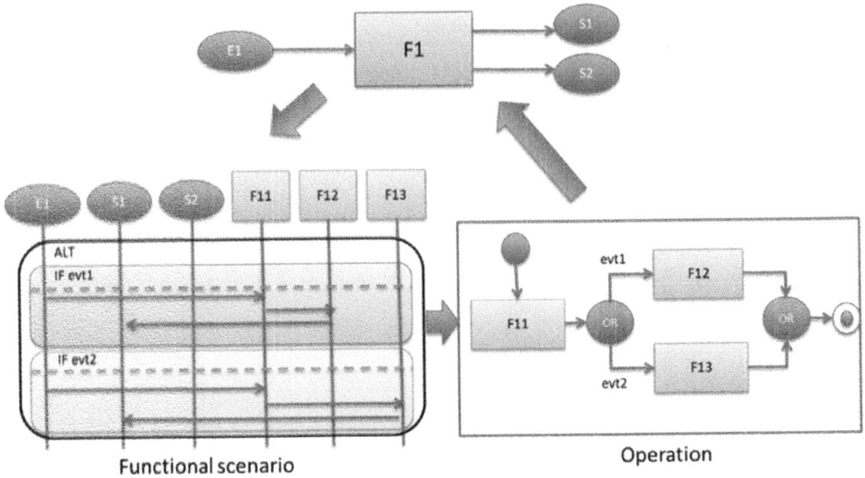

Figure 8.31. *Link between functional scenarios and functions*

With the analysis of events and flows, we can reconstruct the operation of system functions (see Figure 8.31).

8.5.2. *Implementation of the unifying thread example: resumption of the system functioning*

8.5.2.1. *Analysis of the functioning "supply energy from wind"*

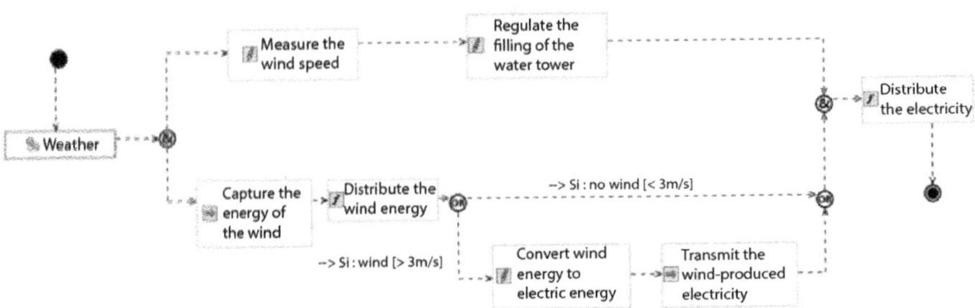

Figure 8.32. *Functioning of "supply energy from wind"*

The functional chain that allows us to provide electrical energy from wind is composed of a system operation function and a control function. The chain of operation of the system function is decomposed into four sub-functions and we have two possible behaviors.

– We can capture the wind energy, and if it is sufficient (wind speed >= 3 m/s), we can distribute it to convert it into electrical energy.

– We can capture the wind energy and if it is insufficient (wind speed < 3 m/s), we do nothing. Indeed, the control system will pass the relay to the electrical network.

To control the two possible behaviors, the system function is decomposed into a chain of two sequential sub-functions, one that measures the wind speed and the other that regulates the filling needs of the tank. We can see in Figure 8.32, the two functional chains in parallel. The first is the control chain that is made up of an acquisition function "measuring the wind speed" and a processing and communication function "regulate the filling of the water tower". The second is the chain of operation with a supply function "capture the wind energy", a distribution function "distribute the wind energy" and if the wind speed is sufficient, we have a conversion function "convert the wind energy into electrical energy" and a transmission function "transmit the wind electricity".

8.5.2.2. *Analysis of the "suck water" functioning*

For the functioning of "route the wind energy", we have two possible behaviors (see Figure 8.33):

– if the tank is empty, the electrical energy of the wind is transported and used to pressurize the water;

– if the tank is full, the generated electrical energy is distributed to the electrical grid after being converted.

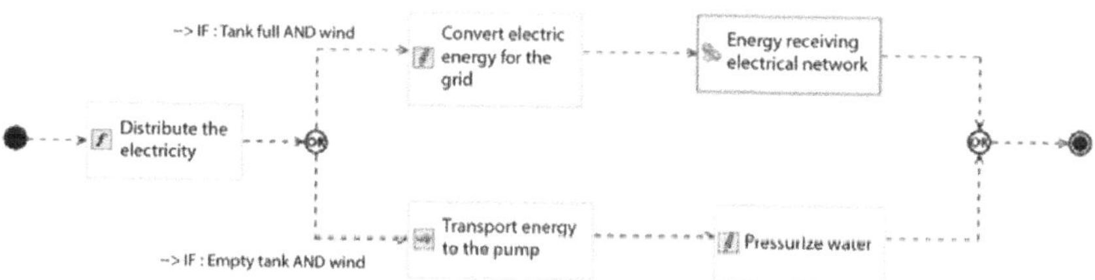

Figure 8.33. *Functioning of "route wind energy"*

8.6. Propose groupings of sub-functions into logical components

8.6.1. *The use of logical components*

The functional decomposition and the physical decomposition of the system are not biunivocal. In order to facilitate allocations, it is interesting to make functional groupings to form logical components. A logical component is a functional module representing a technological principle without being a solution. This results in a more stable architecture that can be easily projected onto different physical solutions.

Let us take the following simple example where we have two functions to perform which are "acquire the temperature" and "acquire the pressure". We allocate these two functions to two logic components that are respectively a temperature sensor and a pressure sensor. Indeed, these two components are closer to the functional abstract view than to the physical concrete view. Now, if we look for solutions to these two logical components, we can have two cases:

– case 1: a solution with a physical sensor combining the two acquisition functions;

– case 2: a solution with two physical sensors, one for temperature acquisition and the other for pressure acquisition.

8.6.2. *Proposing a schematic diagram of the solution*

The schematic diagram of the solution is based on logical components. It is usually identified very early in the process, i.e. at the time of the specification of the requirements. It is usually found in the system presentation. It is made up of high-level components and allows the preliminary organization of the project to be set up.

8.7. Allocate functional requirements to sub-functions

8.7.1. *Objective of the allocation of system requirements to sub-functions*

The objective of this activity is to identify the initial requirements of the sub-systems. Indeed, the requirements allocated to the sub-functions will be allocated by transitivity to the physical components (see Figure 8.34). This system requirements allocation activity is done in parallel with the functional decomposition. In fact, when the system functions are broken down into sub-functions, the associated requirements are broken down and, depending on the case, can be refined or distributed over the sub-functions (see Figure 8.34).

During this analysis and this declination, **requirements induced** by the design choices can appear. We can also talk about requirements derived by design choices (SAE International 2010).

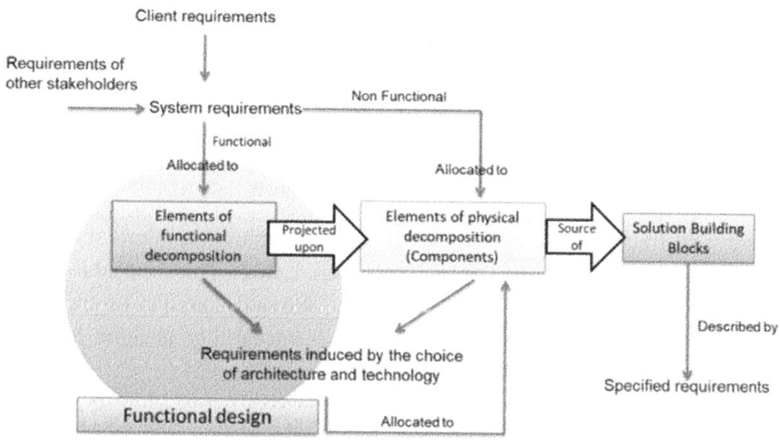

Figure 8.34. *Allocation of functional requirements (ANSI 1999)*

It is mainly the functional requirements of the system that are allocated to the sub-functions. As a reminder, functional requirements are the properties of the system that prescribe what it should do and the associated performance. Since the functional design is done in parallel with the physical design, non-functional requirements can influence the decomposition and allocation.

In Figure 8.35, we have a function F that is decomposed into three sub-functions F1, F2 and F3 and has two requirements Exi1 and Exi2. When allocating, we have the following two cases:

– the requirement Exi1 is declined without refinement to F1 in Exi11;

– the requirement Exi2 is broken down into Exi21 and Exi22 between F2 and F3.

To have a good declination of a system requirement, it is necessary that the conjunction of the declined requirements is more restrictive. For example, for the requirement of Figure 8.35, we have:

$$Exi2 \leq Exi21 \wedge Exi22$$

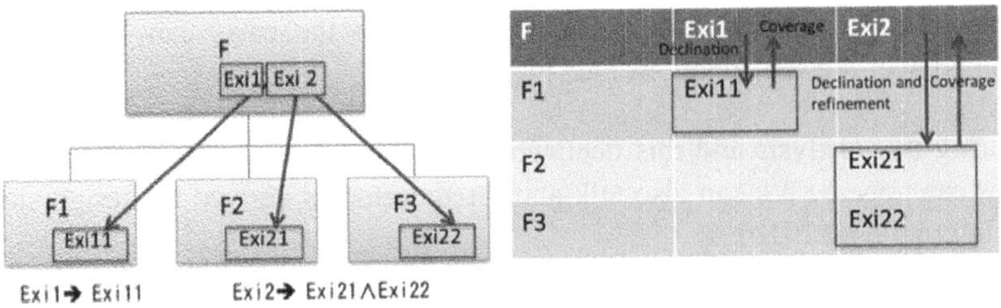

Figure 8.35. *Allocation of functional requirements (AFIS 2012)*

During allocation, we can trace the system requirements to the allocated requirements. There are two points of view in traceability:

– Downward traceability: we will speak of a declination or derivation link, i.e. the requirement Exi2 is declined into Exi21 and Exi22. This traceability is done from an upstream requirement point of view. We are looking to translate this requirement into downstream requirements. This traceability is done at the same time as the writing of the downstream requirements.

– Upward traceability: we will speak of a covering link, i.e. Exi21 and Exi22 cover Exi2. This traceability is done from a downstream requirement point of view. We seek total coverage of the upstream requirement. The downstream requirement, studied for traceability, participates partially or totally in the coverage of the upstream requirement.

8.7.2. *Application to the unifying thread example: allocation of system requirements to sub-functions*

For the application to the unifying thread example, we will start from two system requirements which specify two functions of the system which are respectively "take energy from the air" and "produce electrical energy with the energy of the air".

For the function "take energy from the wind", we have the following requirement:

System Requirement 1:

The system must transform the wind speed into mechanical power between Pmin and Pmax if the wind is between Vmin and Vmax and if the control setpoint is "wind energy" in the "suck water with wind energy" functional mode.

The function “capture wind energy” is broken down into two sub-functions which are “capture wind energy” and “distribute wind energy”. For example, the declination of the system requirement on the sub-function “capture wind energy” could give the following requirement:

Allocated requirement 11:

The function “capture wind energy” must transform the kinetic energy of the wind with a speed between Vmin and Vmax into a mechanical power between Pmin and Pmax in the functional mode “suck water with wind energy”.

We can complete this declination on the sub-function “distribute wind energy” with the following allocated requirement:

Allocated requirement 12:

The “distribute wind energy” function must allow the transmission of mechanical energy if the control setpoint is “wind energy” in the “suck water with wind energy” mode.

For the function “produce electrical energy with air energy”, we have the following system requirement:

System requirement 2:

The system must produce electrical energy between Pelecmin and Pelecmax from the mechanical energy of the wind Pmin and Pmax in “fill the tank and use the wind energy” or “produce electricity for the grid” mode.

The sub-function “convert wind energy into electrical energy” is derived from the decomposition of the system function “produce electrical energy with wind energy”. Declining this requirement on this sub-function could result in the following allocated requirement:

Allocated requirement 21:

The function “convert wind energy into electrical energy” has to transform the mechanical energy between Pmin and Pmax into electrical energy between Pelecmin and Pelecmax in the mode “fill the tank and use the wind energy” or “produce electricity for the grid”.

As a rule, the functional and physical designs are done in parallel. We could have allocated the requirements directly to the component architecture. We would then have replaced "the sub-function must" with "the component must".

We can note that if the component is logical, we are in the case of an initial sub-system requirement and if the component is physical, we may be in the case of a specified requirement, i.e. a requirement used to realize the system.

8.8. Conclusion on functional design

Functional design is a pivotal process between the specification of the system that poses the problem to be solved and the physical design that gives the solution that must be realized. To do this, the functional design process proposes to study the white-box system from an abstract perspective. This design results in a functional architecture by decomposing the system into sub-functions from a static, discrete event dynamic and behavioral (continuous dynamic) perspective.

For the static part, the approach proposes to decompose the system functions into sub-functions and to detail the network of internal and external flows. For the discrete event dynamics part, we study two aspects. First, we consolidate the view of the operating modes and their events. Then, the approach proposes to study the functions of the system under an algorithmic angle. For the behavioral part, we synthesize the functional modes and sub-functions in order to render the behavior of the functions and missions of the system in the form of functional scenarios.

In parallel, the functional design activity consists of declining the requirements of the system on the sub-functions.

9

Verification and Validation of the Functional Architecture

9.1. Verification of the functional architecture

In general, verification of functional architecture consists of ensuring the conformity of the implementation of the functional design process (AFIS n.d.).

The functional architecture verification activity involves checking the following points (Faisandier 2013):

– the allocation of functional requirements;

– the coherence of the sub-functions;

– the coherence of the system dynamics;

– control and behavior of the sub-functions;

– sub-functions or logical components;

– the interactions.

From a **static angle**, we need to check that:

– each sub-function has consistent input and output streams;

– each sub-function produces at least one output;

each sub-function covers at least one functional requirement.

We must also check that each functional requirement is allocated to at least one sub-function.

Systems Engineering,
by Emmanuel HYGOUNENC. © ISTE Ltd 2023.

From a **dynamic perspective**, we need to verify that each sub-function is activated at least once in a functional mode after an event.

From a **behavioral perspective**, we need to verify that:

– each sub-function participates in the functioning of a functional chain or a functional scenario;

– the control of the sub-functions, which must be regulated, is consistent with the expected behavior.

From a **property perspective**, each sub-function must be specified from the allocation of system requirements or design-induced requirements. Interactions between sub-functions must be characterized.

9.2. Validation of functional architecture

In general, validation of the functional architecture involves ensuring that the repository of requirements allocated to the sub-functions covers the system requirements repository.

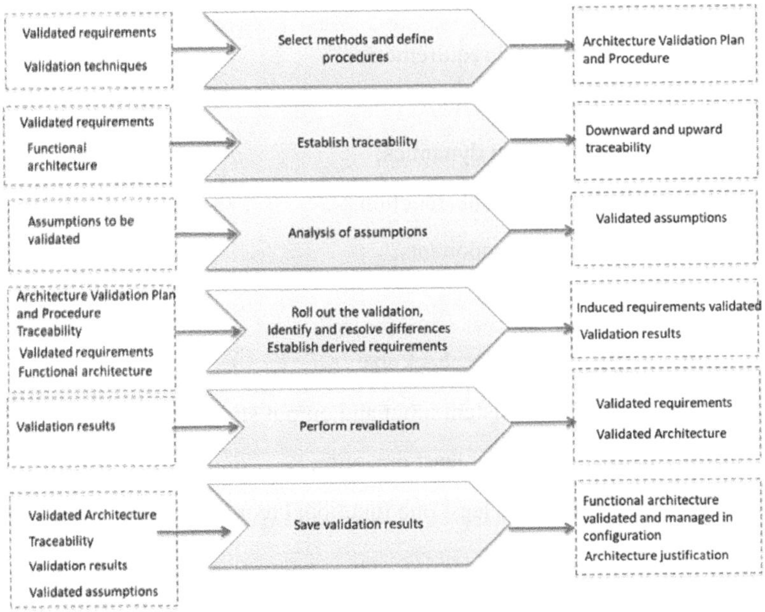

Figure 9.1. *Validation of the functional architecture. For a color version of this figure, see www.iste.co.uk/hygounenc/system.zip*

To conduct this activity, the following points are validated:

– the sub-functions respond well to the functions of the system;

– allocated requirements meet system requirements;

– the activation dynamics of the sub-functions correspond well to the activation dynamics of the system functions;

– the behavior of the sub-functions covers the operational scenarios of the system well.

To carry out this validation of the functional architecture, we can work through the process in Figure 9.1, which is inspired by the EIA-632:1999 standard (ANSI 1999).

9.2.1. *Selecting methods and defining validation procedures*

First, we need to select and define the methods for validating the functional architecture and derived requirements. This definition of methods must be done in accordance with the system validation plan.

9.2.2. *Establishing traceability*

We need to establish bidirectional traceability between the system requirements repository and the functional architecture with its derived requirements.

We need to do top-down traceability between the validated system requirements and the functional architecture elements with their derived requirements.

We need to establish the traceability between the functional architecture elements with their requirements and the validated system requirements.

9.2.3. *Analysis of assumptions*

Throughout the functional design, assumptions may have been made to consistently meet the system requirements. These assumptions must be evaluated to determine whether they are valid. Among the types of assumptions, we can have:

– performance on external systems (to be raised with relevant stakeholders);

– technological performance, which results in risky requirements on a subsystem;

– the understanding of a requirement (to be validated with the corresponding stakeholder).

9.2.4. *Unrolling the validation*

The validation is carried out according to the methods and procedures that have been selected and defined. Identified gaps or conflicts between the system requirements and the functional architecture with its derived requirements must be resolved. If a system requirement is not covered or is poorly covered by the functional architecture and its derived requirements then the functional architecture must be corrected. On the contrary, if elements of the functional architecture do not cover any of the system requirements, then a request for changes to the requirements repository must be made.

9.2.5. *Performing the revalidation*

This revalidation step concerns both the functional design and the technical requirements repository. Indeed, if a request for a change in the requirements repository has been issued, then the added or modified elements must be verified and validated. All or part of the repository must be revalidated to avoid regressions or the appearance of gaps or conflicts. The functional architecture elements that have been added or modified must be verified and validated. In this case, the functional architecture must be partially revalidated.

9.2.6. *Recording the results of the functional architecture's validation*

The functional architecture must be versioned and managed in the configuration. The validation procedures with their results, as well as the assumptions and the different corrective actions must be recorded. They are part of the system architecture justification file.

9.3. Conclusion on the verification and validation of functional design

All systems engineering activities include verification and validation activities. Figure 9.2 shows the architecture interactions with other engineering activities. For the validation of the functional design process, we will make sure that the functional decomposition meets the functions and missions of the system. First, it is necessary to verify that the functional architecture has been created according to the rules of the art.

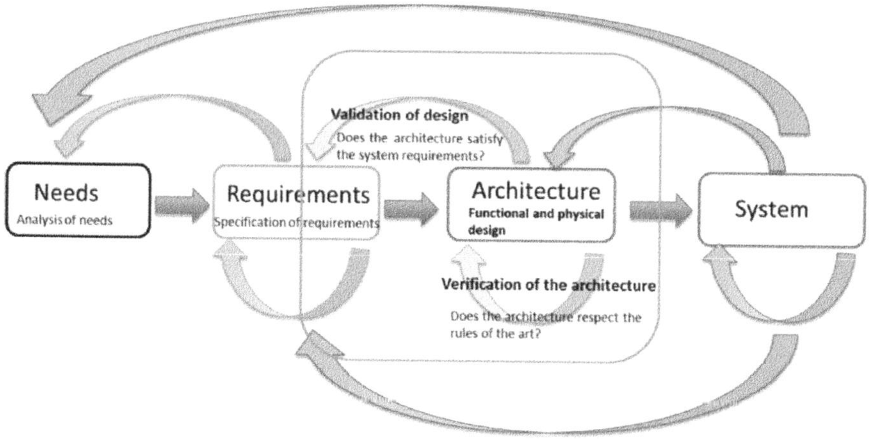

Figure 9.2. *Verification and validation of the functional architecture. For a color version of this figure, see www.iste.co.uk/hygounenc/system.zip*

10

Physical Design

10.1. Purpose of physical design

The objective of the physical design process is to identify one or more architectures that can be realized, addressing the problem posed by the system specification. The physical architecture resulting from this design process must implement the functions of the functional architecture and meet the missions of the operational architecture. This physical architecture will be defined by specified requirements of subsystems that will allow the realization of the system. This definition will also be used for the integration of the system and for the rest of its life cycle, such as its commissioning and maintenance.

As with the functional design process, the physical design process will study the architecture along a threefold dimension (see Figure 10.1):

– static with a component tree and interface analysis;

– dynamic with the identification of the technical configurations of the system and the associated events;

– behavioral, which is the integration of the two previous views with the objective of reconstructing the functional and operational scenarios of the system.

However, this physical process is a **synthesis process** which is articulated around a dual top-down and bottom-up approach (see Figure 10.2). This top-down approach consists of a set of engineering activities, which take as input the outputs of the operational and functional architectures as well as the physical properties sought, i.e. the set of non-functional requirements that will influence or impose some of the system's components. This upward approach is made up of a set of integration activities, which take into account the company's technological

---

For a color version of all the figures in this chapter, see www.iste.co.uk/hygounenc/system.zip.

Systems Engineering,
by Emmanuel HYGOUNENC. © ISTE Ltd 2023.

possibilities or the state of the art in the field (available solutions, existing technologies), the market solutions, as well as the company's innovation strategy.

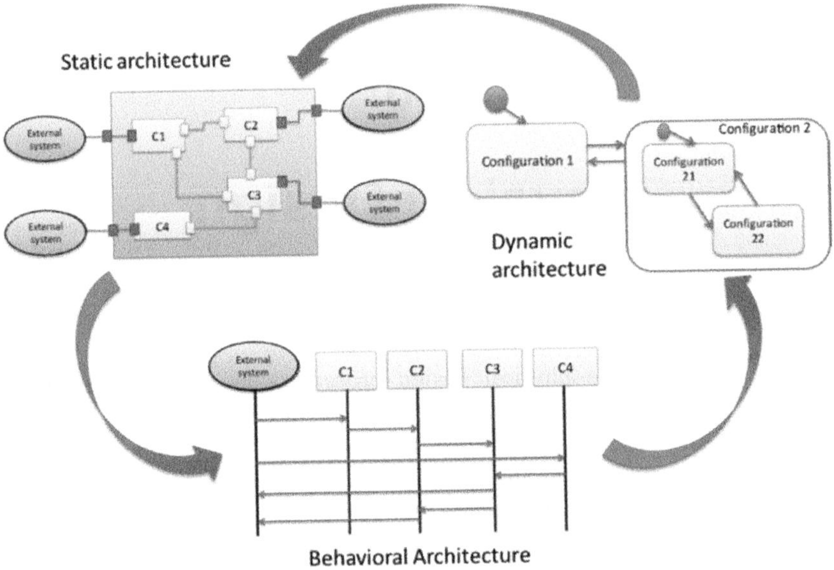

Figure 10.1. *Three crossed visions of the physical architecture*

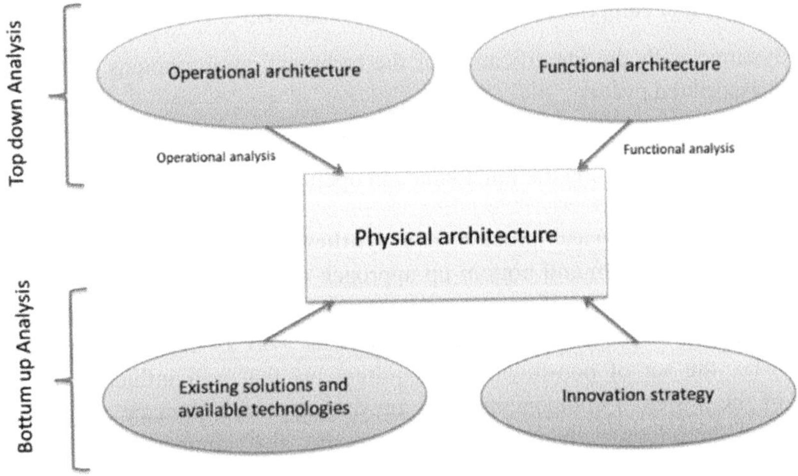

Figure 10.2. *Physical design, a synthesis process*

10.2. Identification of physical components

As input to this activity, we have the leaf sub-functions of functional design, the system missions of operational analysis and non-functional requirements such as design requirements.

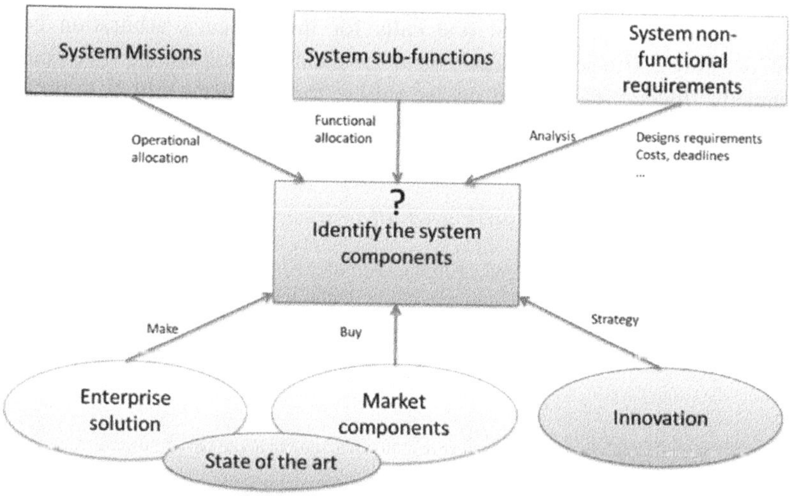

Figure 10.3. *Identifying the system components*

We also have as input the company's know-how with the available technologies, the state of the art as well as the innovation strategy (see Figure 10.3). At the end of this activity, we will have a list of components with possible choices to arbitrate and prioritize. A component of a system is a concrete element of this system which can be:

– material, such as a physical component;

– computer science, such as software or its "hardware";

– human, such as an operator or an organization.

A component can be elementary or the result of the integration of other components. The identification of the system components is based on:

– the sub-functions to be allocated from the functional architecture;

– the missions of the system;

– non-functional system requirements that influence the choice of components.

On the other hand, the identification of components is also based on:

– the company's know-how and market solutions;

– risk-taking in business innovation.

These last two points have an influence on decision-making regarding the choice of constituents. Indeed, the first one calls for the project's arbitration between "Make" or "Buy". The second depends on the technical and competitive context of the company. Indeed, depending on its policy, the company will have to choose between **disruptive innovation** and **incremental innovation**.

10.2.1. *Identifying components and allocating technical functions*

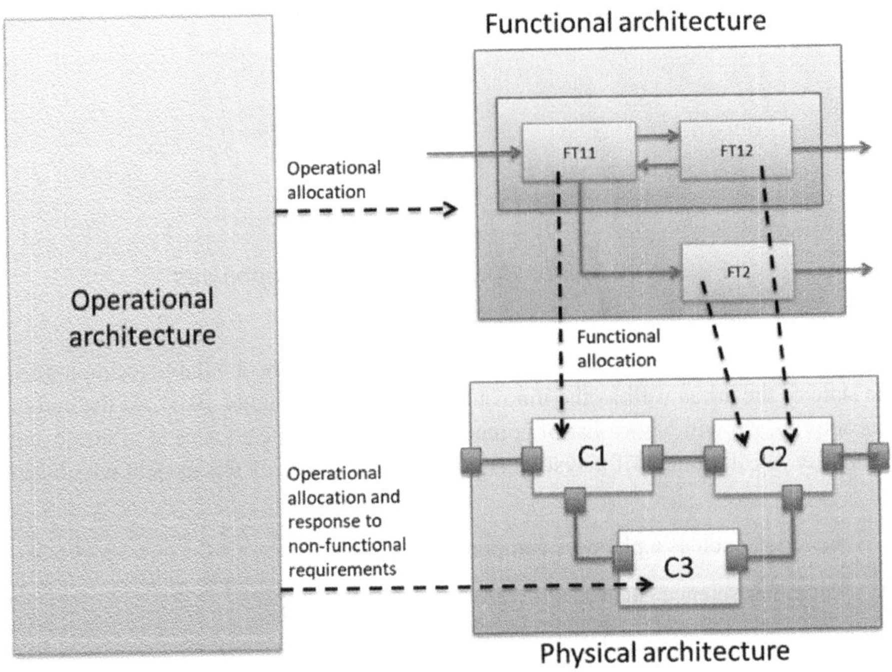

Figure 10.4. *Functional and operational allocation*

For each leaf function resulting from the functional decomposition of the design process, we will try to identify one or more candidate components that can implement it with the associated performances (see Figure 10.4).

10.2.2. *Completing the identification of the components with the missions of the system*

The analysis of the system's missions makes it possible to identify new components and to consolidate the properties of the components already identified through the functional allocation. To do this, it is sufficient to take up the different missions of the system and to analyze for each of them their constructional dimension. That is, we try to answer the mission not with functions, but directly with components of the system.

10.2.3. *Completing the identification of components with non-functional requirements*

The analysis of the non-functional requirements of the system allows for identifying new components or to consolidate the properties of the components already identified through the functional and operational allocation. To do this, we simply select non-functional requirements, such as design requirements, and analyze them in the form of a constructional response scenario, as proposed by Krob (2017) in the CESAM approach.

10.2.4. *Identification of variants in the physical architecture*

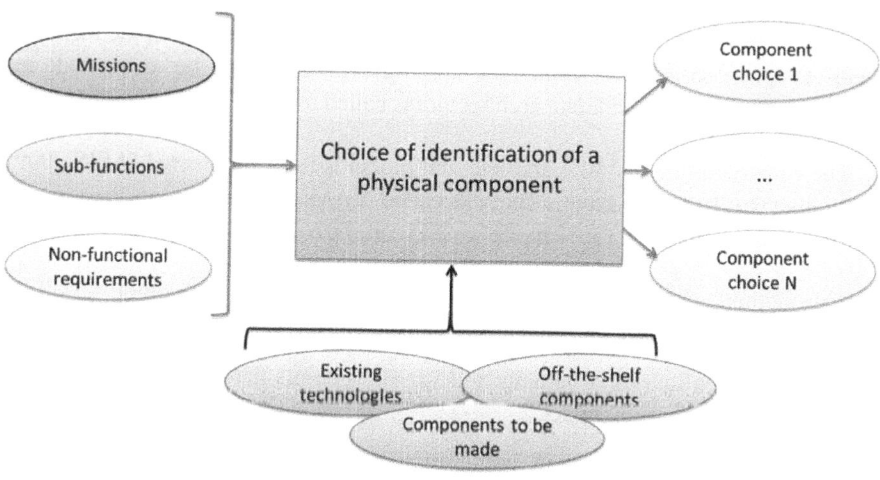

Figure 10.5. *Choice of components*

During the identification activity, several components can implement the same functions and meet the constraints imposed by the non-functional requirements (see Figure 10.5). In systems engineering, the process of evaluating and comparing architectures is a high-value process. Indeed, it avoids the project starting on the development of the first solution found which is perhaps not the best. It also gives the project the opportunity to compare several candidate architectures and choose the one with the best compromise.

At this point in the solution design process, keep all possible paths to solutions open. For each sub-function, we identify the possible component choices. We do the same by completing the identification activity with the analysis of the missions and non-functional requirements.

10.2.5. *Building physical architectures with a product line approach*

This approach enables the selection of candidate architectures from a component base consisting of two parts:

– a core architecture with the list of mandatory components;

– a variant architecture with the list of components to be selected and the associated constraints.

This base of constituents is called a 150% architecture. It is then possible with this type of architecture to cover a wide operational spectrum and from the identified needs to propose several candidate architectures that can be evaluated and compared. Each 100% candidate architecture is called a configuration variant.

The construction of a 150% architectural model is fully engineered (AFIS 2009). In Weilkiens (2014), we have a method called VAMOS that presents in a simple way the construction of a 150% model (see Figure 10.6).

This method is based on three models that are connected to each other:

– the "CORE" model contains an architecture consisting of core elements that are part of all configuration variants and variation points that are dummy elements connected to the variation model;

– the "VARIATIONS" model contains the variations that are connected to the "variation points" and that contain the variants, as well as all the constraints associated with these variants;

– the "VARIANT CONFIGURATION" model, which is a 100% architecture model consisting of the "CORE" and a selection of variants that respect the constraints imposed by the "VARIATIONS" model.

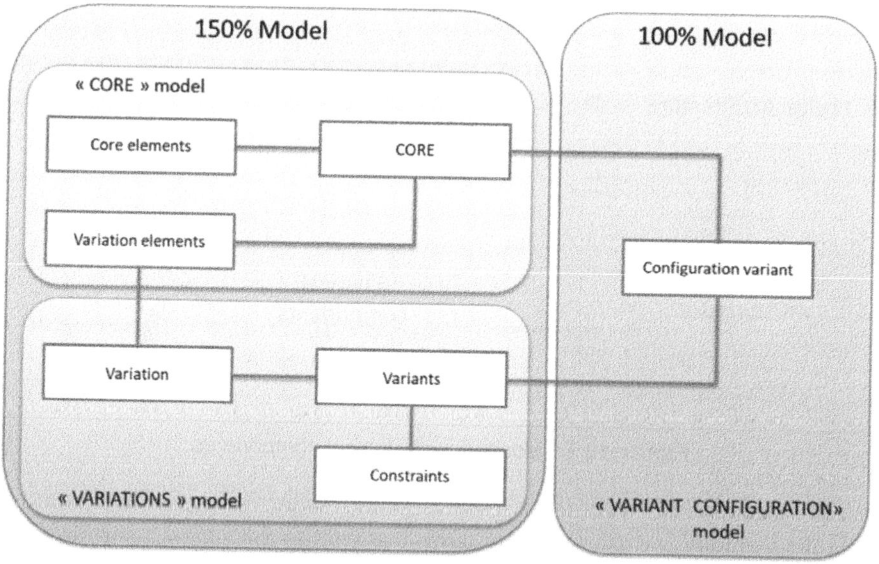

Figure 10.6. *150% model*

In order to build a 100% model from a 150% model, it will be necessary to identify constraints between the variants which are of two types:

– the "XOR" constraint, i.e. the "Exclusive OR": two variants that have an XOR constraint cannot be part of the same configuration variant;

– the "REQUIRES" constraint: two variants that have a REQUIRED constraint are necessarily part of the same configuration variant if they are selected.

10.2.6. *Building physical architectures from a logical architecture*

The functional decomposition and the physical decomposition are not biunivocal. In order to facilitate functional allocation and the identification of possible physical component choices, it may be interesting to use a logical component architecture which is more stable and more easily allocated to physical solutions.

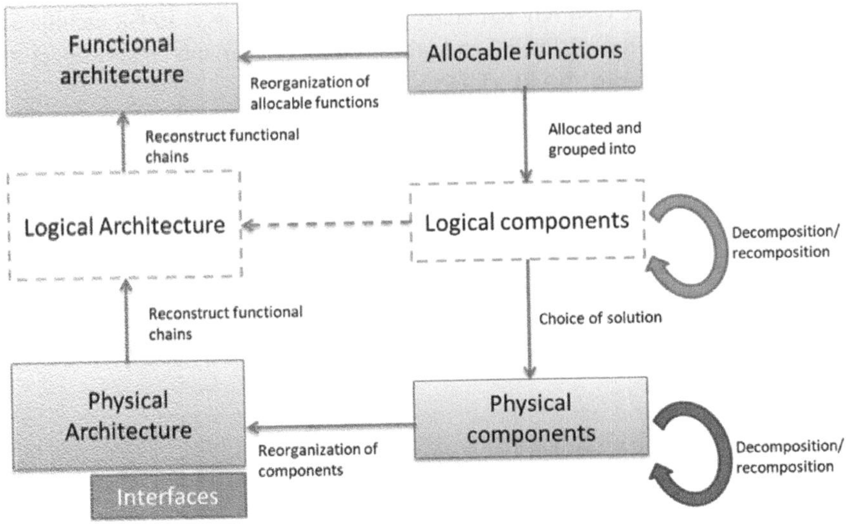

Figure 10.7. *Allocation with logical components*

In the literature, we can find other terms that refer to the logical component like the functional module (AFIS 2012) or the constructional component (Meinadier 1998).

In Figure 10.7, the allocatable functions are grouped into logical components which are allocated to one or more physical components. The functional architecture is projected onto the logical architecture, which, in turn, is projected onto one or more physical architectures.

For example, we have an allocatable function that is "measure temperature". This function is allocated to a logical component which is an "electric dipole temperature sensor". This "electric dipole temperature sensor" can be allocated to a "thermistor" or a "PT1000 thermo probe" which are two possible choices of physical components. Logic components can represent a technological way with, for example, constraints on interfaces.

10.2.7. *Application to the unifying thread example: identification of components*

In our unifying thread example, the analysis of the sub-functions allowed us to identify the following physical components (see Figure 10.8).

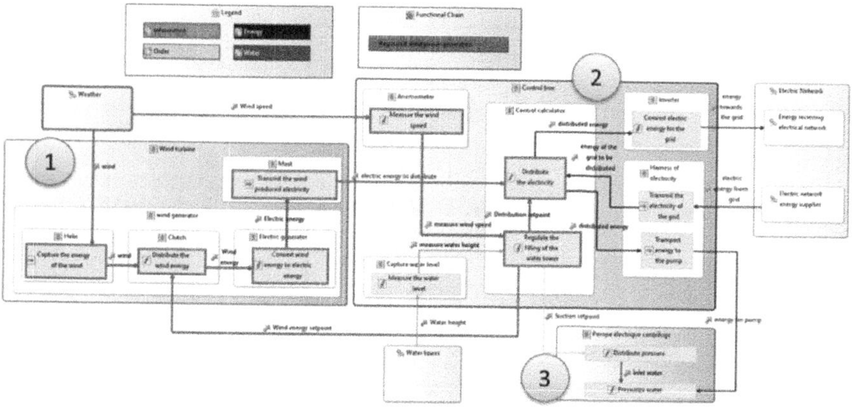

Figure 10.8. *List of components of the unifying thread example*

The physical components are hierarchically organized into three subsystems, which are the "wind turbine" (see Figure 10.9), the "control box" (see Figure 10.10) and the "electric centrifugal pump" (see Figure 10.11).

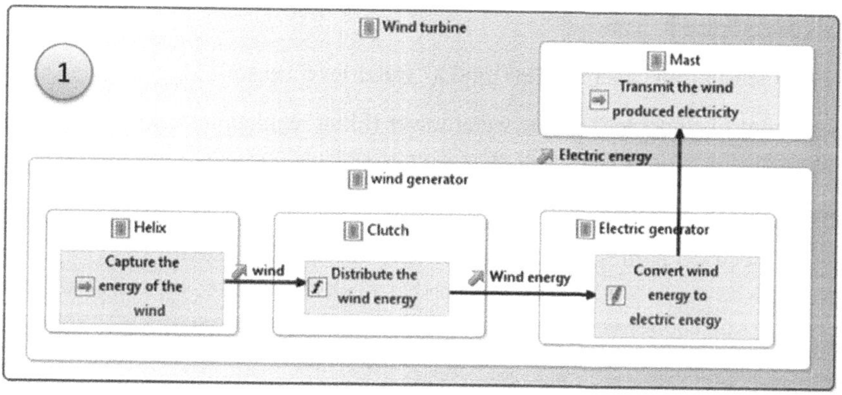

Figure 10.9. *Zoom on the "Wind turbine" component*

Figure 10.9 presents the subsystem "wind turbine" with two pieces of equipment which are the "mast" and the "wind generator". The wind generator is broken down into three components which are the "propeller", the "clutch" and the "electric generator".

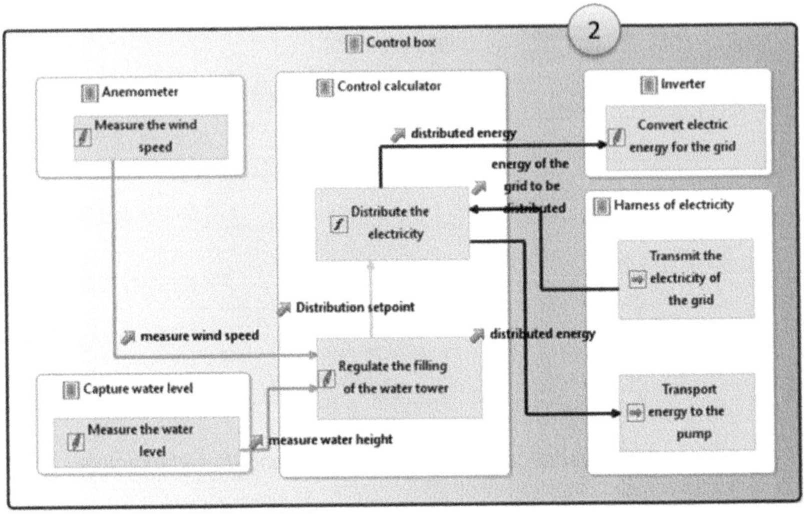

Figure 10.10. *Zoom of the "control box" component*

Figure 10.10 shows the decomposition of the "control box" subsystem into four pieces of equipment, namely:

– two sensors, an "anemometer" and a "water level sensor";

– a "control computer" for the water tower filling regulation treatments;

– an "inverter" for the respect of the constraints of injection of the electricity produced on the network;

– an "electrical harness" to carry electricity through the system.

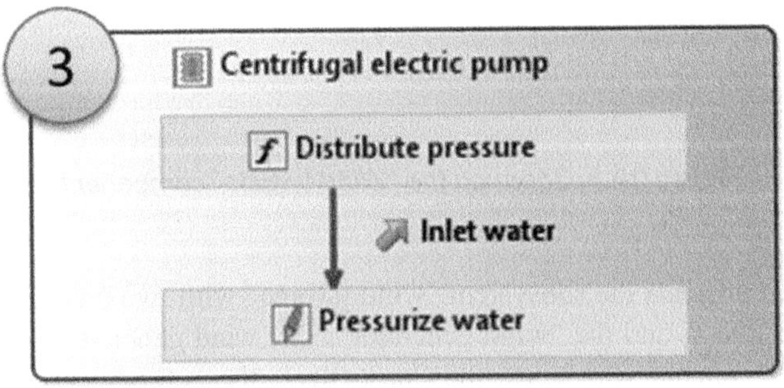

Figure 10.11. *Zoom on the "electric centrifugal pump" component*

Figure 10.11 shows the subsystem “electric centrifugal pump” with the interface function, with the control “distribute pressure” and the conversion function “pressurize water”.

10.3. Decomposition of the components

The objective of this activity is to group the identified components to form the product tree, in order to allow for identification of the physical interfaces and the dimensioning of the solution (see Figure 10.12).

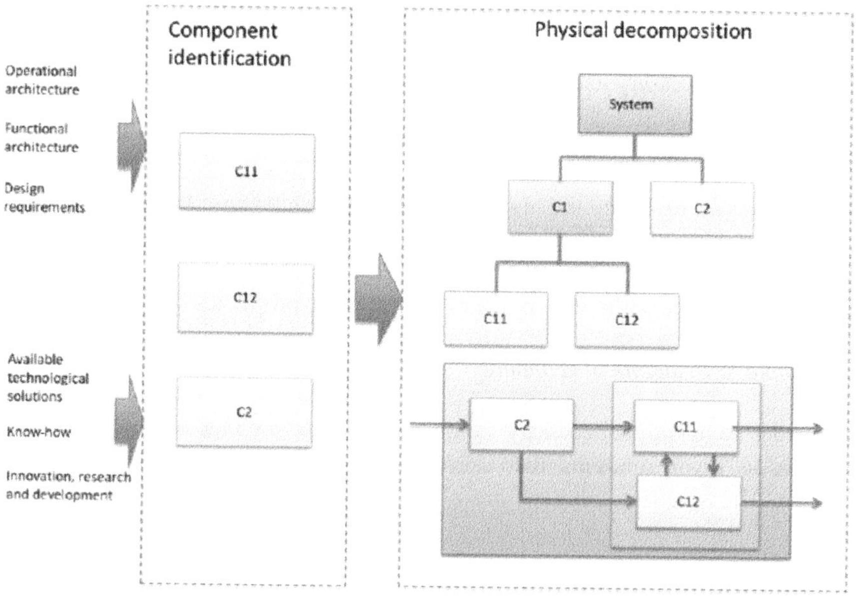

Figure 10.12. *Physical decomposition*

As an input to this decomposition activity, we have the list of hardware, software and human components that have been selected to constitute a candidate solution. In output, we will have a product tree with the identification of the subsystems and the physical architecture with the interfaces to be characterized. To do this activity, the engineer can rely on architecture principles and coupling analysis tools such as the DSM, “Design Structure Matrix”.

10.3.1. *Architecture principles*

To build a good physical decomposition, we can rely on some architectural principles stated in Krob (2017). In addition to the principles proposed by Krob, we can add the principles of systemic breakdown of Donnadieu and Karsky (2002).

Principle 1: Optimization principle

This principle is based on the importance of clearly defining the architectural problem to be solved. The physical architecture is an optimization problem for which the criteria to be minimized and maximized must be clearly identified. To do this, the criteria to be optimized must be defined with their integration and optimization law. In order to maintain technical feasibility and avoid getting lost in the details, it is necessary to prioritize these criteria.

Principle 2: Stabilization principle

It is important to secure the high-level physical architecture. Indeed, to optimize this architecture problem, it is necessary to stabilize the non-functional system level properties (such as dimensions, mass, reliability, availability). Indeed, there is no point working on the optimization of subsystems if the properties of the system vary.

Principle 3: Principle of economy

In order to avoid unnecessary complexity of the architectural solution, it is preferable to favor solutions that use the least amount of resources (technical, financial). The simplest solutions are always the best, as they are often the most robust.

Principle 4: Principle of symmetry

The idea of this principle is to organize the components along an axis of symmetry. This axis of symmetry can be physical or informational. It is important to identify the axes of symmetry early in the physical design process. For example:

– for the physical axes, we can have a geometric symmetry, a balance of masses, of consumption or production of energy;

– for the informational axes, we can have a balance of transmission and reception of data or a balance of transmission and reception of information flows.

Principle 5: Decoupling principle

In order to facilitate system integration and limit failure propagation, it is recommended to have strong internal cohesion of each component and weak coupling between components by minimizing interactions (Meinadier 1998). Minimizing the interactions facilitates the solution of the integration problem. To do this component grouping analysis, we can use the coupling matrices (Eppinger and Browning 2012). To evaluate the best decomposition/grouping, we can use the **global coupling quality criterion** (Faisandier 2013). Decoupling can also be done at the level of the functional architecture by being careful not to allocate functions to the same component.

Principle 6: Principle of ergonomics

This principle, based on the 7x7x7 rule, recommends breaking down the system into a maximum of seven elements, which are themselves broken down into seven sub-elements, which are themselves broken down into seven sub-sub-elements. An element can be a component or a function. Without seeking a strict application of this rule, the objective is to start from the fact that it is useless to share an incomprehensible architecture. It is a rule of alignment between the product and the project in order to facilitate the sharing of the envisaged solutions.

Principle 7: Principle of finality

This principle is based on the identification of the function of the subsystem in relation to the whole, i.e. the system, the other subsystems and the external systems.

Principle 8: Historical principle

This principle consists of studying the dynamics of subsystems and identifying the behaviors they have in common. According to the CESAM approach, this consists of identifying the technical configurations of the system (Krob 2017).

Principle 9: Organizing principle

This principle is based on the levels of modeling of a complex system proposed by Le Moigne (1999). The idea is to identify where the studied subsystem is located in relation to the hierarchical organization of the system according to its level of complexity (see Figure 10.13).

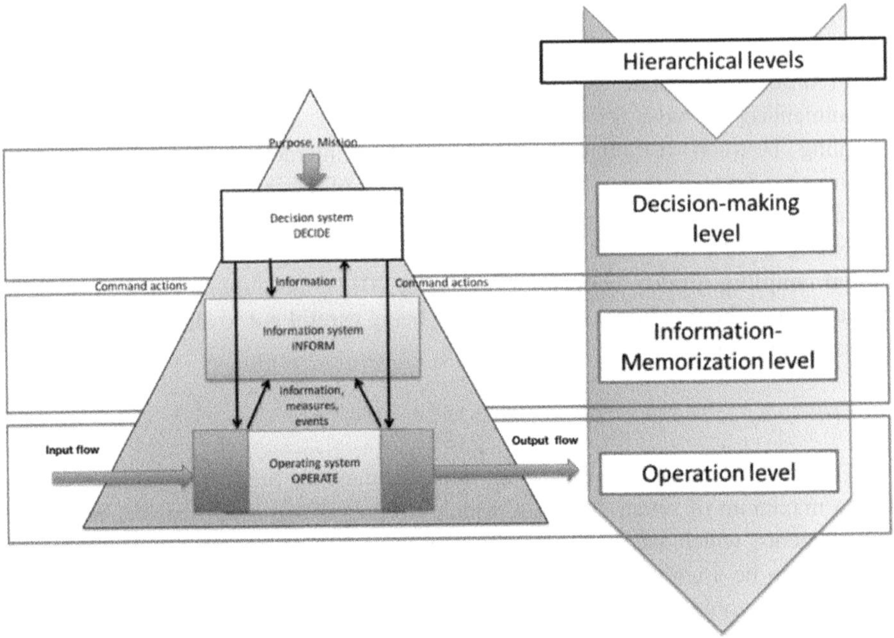

Figure 10.13. *Hierarchical level on an O.I.D model*

Principle 10: Principle of structure

This principle is based on the fact that there are repeating structures in systems that can be found on several levels of organization. These repetitive structures are called **fractal** or **hologram structures**. The modeling of one of these holograms allows us to know them all.

Principle 11: Principle of ownership

This principle, linked to project management, recommends being able to associate an architectural element, i.e. a function or component, with a person or an organization.

Principle 12: Separation principle

This principle implies that only data under the responsibility of the owner of the element (function or component) can be defined and modified locally. The others must be provided.

10.3.2. *Coupling matrix*

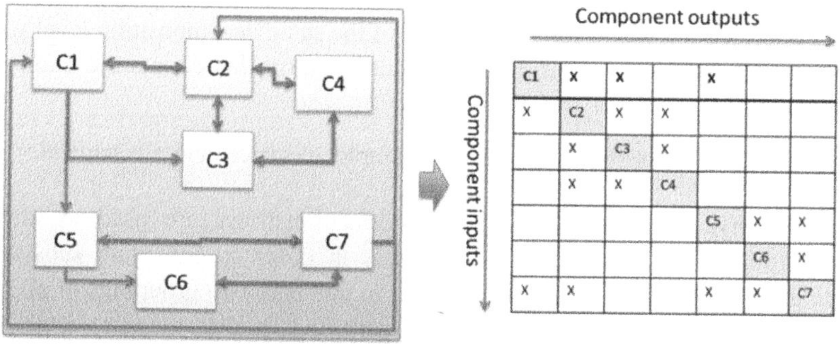

Figure 10.14. *Coupling matrix*

The coupling matrix, also called the N2 matrix or the DSM (Design Structure Matrix) (Eppinger and Browning 2012), is a matrix modeling tool to visualize the interfaces of a physical architecture. This approach is the same as for the functional architecture, but with different objectives.

In Figure 10.14, we have a Boolean matrix that displays the presence or absence of a flow between the components. The construction of this matrix is simple:

– in each row and column, we list the components of the studied level;

– each row of the matrix represents the outputs of the selected component;

– each column of the matrix corresponds to the entries of the selected component;

– the row/column intersection corresponds to an interface between the row component and the column component;

– for this type of DSM, we put a cross (or 1 or TRUE) if there is at least one flow exchanged between the two components.

This puts in matrix form the flows or interface links of a physical architecture. We can represent in the form of a DSM:

– the flows exchanged between the components of a system **decomposition level** with a binary or Boolean DSM that displays the presence or absence of an interaction or a DSM that counts the number of flows between the components;

– the flows exchanged and classified by type of flow with the addition of spatial interactions; this DSM is called a four-cell coupling matrix;

– the interface links between components of the same decomposition level with a binary matrix (interaction or no interaction) or a matrix that counts the number of interactions;

– interface links classified by type, for example, electrical and mechanical.

For the physical architecture, these interface matrices are used to identify relevant groupings of elementary components in order to constitute subsystems respecting the **decoupling principle**. In order to compare two groupings of physical architecture into subsystems, we can use a **coupling quality** measure established from DSMs.

10.3.3. *Global coupling quality of an architecture*

Among the criteria for comparing two candidate physical architectures, we have the **quality of their coupling**. The more elementary components there are in a system, the higher the combinatorial range of possible groupings into subsystems. In order to objectively compare two types of clustering, we can establish a quality measure of the couplings. This mathematical tool, which is based on DSMs, provides a score that will highlight the decompositions that have the best internal flow of subsystems and the least external coupling of subsystems to each other. For this purpose, we define the **coupling ratio**:

$$R_i = \frac{Number\ of\ couplings}{Number\ of\ possible\ couplings}$$

We will call a coupling a set of oriented interactions between two components.

The **coupling quality** measurement of the subsystem is established as follows:

$$S = R_{i\ internal} - R_{i\ external}$$

with:

– $R_{i\ internal}$ the internal coupling ratio of the subsystem, i.e. the number of couplings between the components of the subsystem over the number of possible couplings;

– $R_{i\ external}$ the external coupling ratio of the subsystem, i.e. the number of couplings between the subsystems over the number of possible couplings.

The coupling quality is between -1 and 1 where:

– The value -1 corresponds to a coupling with **no internal cohesion** and a **strong external cohesion**. This value corresponds to the worst possible grouping.

– The value 1 corresponds to a **strong internal coupling** and **no external coupling**. This clustering seems ideal, but it is of no interest because the cluster no longer interacts with the environment. It is a closed system.

– A negative value means that the cluster has more external coupling than internal coupling. This configuration is not ideal but not catastrophic as long as it remains far from -1.

– A positive value will mean that the cluster has more internal than external coupling. This configuration is interesting as long as it remains far from 1, in which case it would lose all its interest.

To compare several architectural groupings, we need to calculate the quality of the global coupling which is the average of the couplings of the subsystems:

$$C_s = average\ of\ the\ coupling\ qualities\ of\ the\ subsystems$$

We can test this approach on an example, that is, a system made up of seven components with numerous interactions. We propose two possible decompositions for evaluation.

In Figure 10.15, we have a first grouping of our seven components in three heterogeneous clusters with the S1 cluster with only one component C1, the S2 cluster with three components C2, C3, C4 and the S3 cluster also with three components which are C5, C6 and C7. The overall coupling rate of this architecture is 0.71. It is a positive coupling close to 1 which means that the system has subsystems with strong internal cohesion and with little external coupling.

In Figure 10.16, we have a second grouping of our seven components into three heterogeneous clusters with an S1 cluster with two components C1 and C5, an S2 cluster with three components C2, C3, C6 and a final S3 cluster with two components C4 and C7. The global coupling rate of this architecture is -0.16. It is a negative coupling which means that the internal cohesion is weaker than the external cohesion. This architecture will be more complex to integrate as it has more interaction between the subsystems.

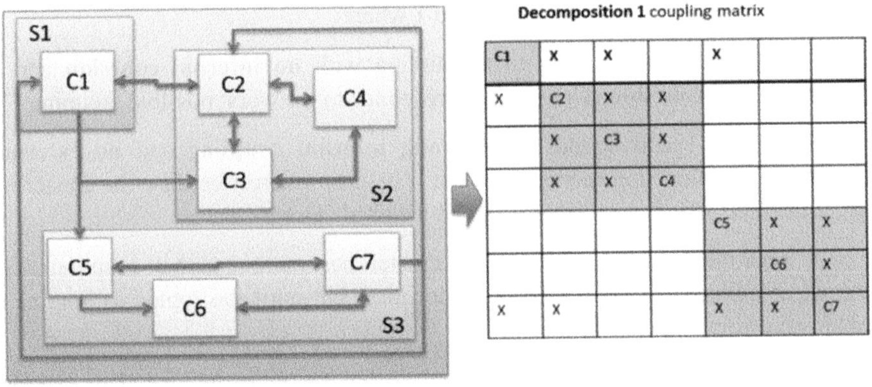

S1 = 1 – 5/12 =0.58

S2= 6/6 – 4/24 =0.83

S3= 5/6 – 3/24 =0.71

Cs = (0.58+0.83+0.708)/3 = **0.71**

Figure 10.15. *Coupling rate for decomposition 1*

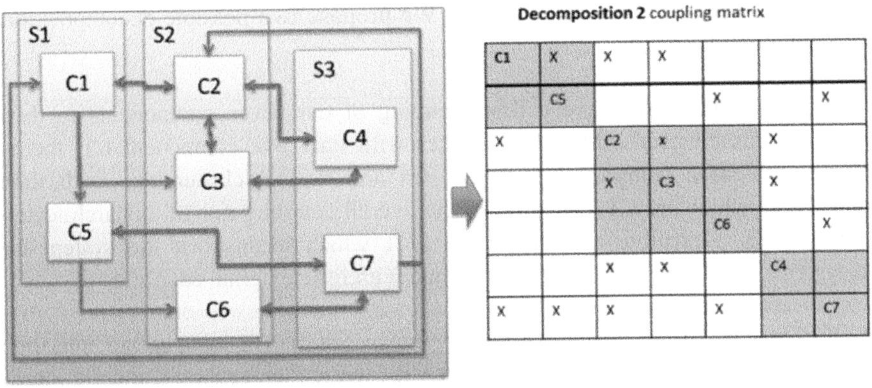

S1 = 1/2 – 4/20 =0.15

S2= 2/6 – 11/24 =-0.12

S3= 0/2 – 10/20 =-0.5

Cs = (0.15-0.12-0.5)/3 = **-0.16**

Figure 10.16. *Coupling rate for decomposition 2*

The better of the two architectures is therefore the first. This approach allows us to deduce it on the basis of an objective and formal criterion.

10.3.4. *The physical tree structure*

10.3.4.1. *The PBS for product breakdown structure*

The physical tree is the decomposition of the system into hierarchical components in order to:

– achieve realization within the company and by the suppliers of the project;

– achieve integration, verification and validation.

The system tree is called the PBS for "product breakdown structure". An example of a preliminary physical cutout for a civilian aerial imaging drone is shown in Figure 10.17.

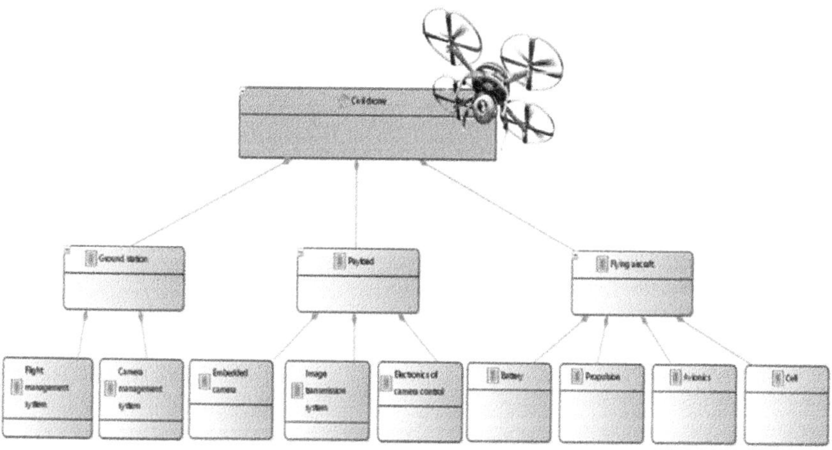

Figure 10.17. *PBS of a civil drone*

This decomposition is made early on in the project, because it allows for initial organization of the tasks, the WBS (workpackage breakdown structure), as well as an initial estimate of the costs and the deadlines (Galinier et al. 2017).

10.3.4.2. *Application to the unifying thread example: the PBS*

Figure 10.18 shows the constructional division of our unifying thread example on four levels. We have the first level which is the integrated "renewable energy water pump" system.

At the second level, we have the subsystems with the centrifugal pump, the control box, the wind turbine and the transport boxes.

At the third level, we have the equipment. For the wind turbine which is decomposed in a mast and a wind generator, a fourth level was necessary which is a component level.

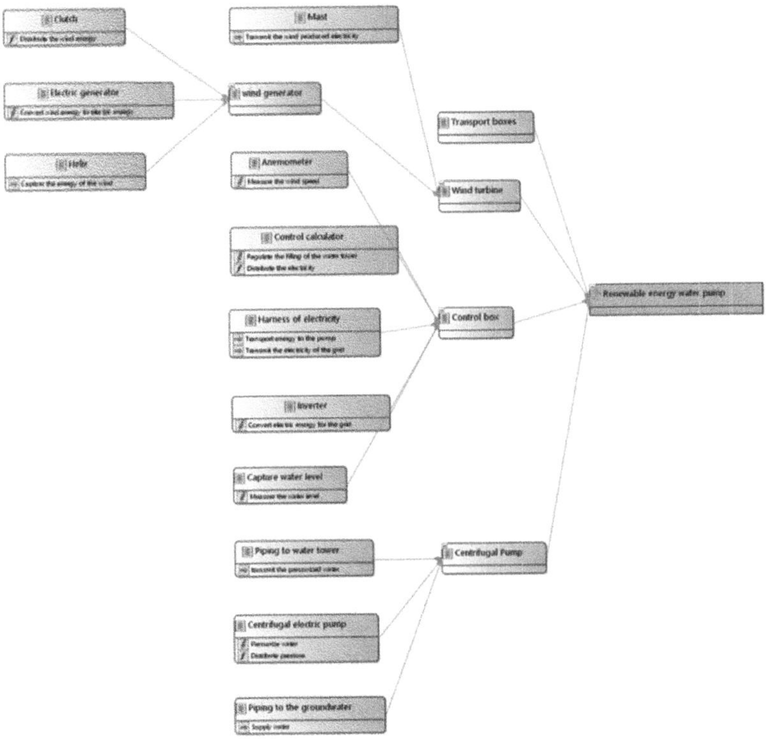

Figure 10.18. *PBS of the unifying thread example*

The PBS of Figure 10.18 shows the breakdown of the system on four levels with the technical functions allocated to the lowest level components. Figure 10.19

presents another formatting of the PBS of our unifying thread example with the decomposition of each subsystem.

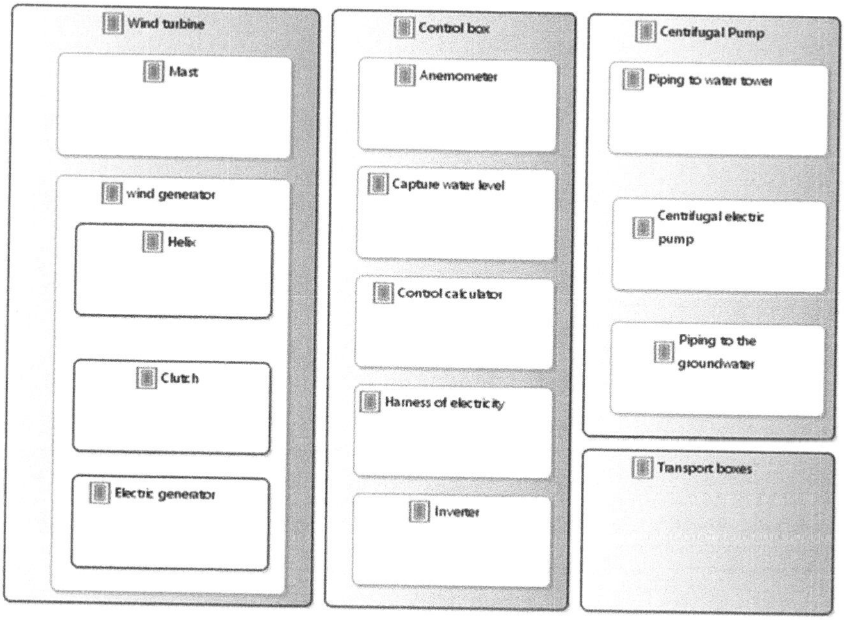

Figure 10.19. *Other formatting of the PBS unifying thread example*

10.3.5. *The architecture of the components*

This step in the design of the physical architecture consists of identifying the interfaces between the components that carry the MEI flows (material, energy, information).

The functional allocation identified the MEI flows that are exchanged between the components that perform the sub-functions, as shown in Figure 10.20. From this physical architecture with MEI flows, we need to:

– identify the physical links that connect the components together;

– identify the physical interface ports that allow each component to be autonomous and interchangeable.

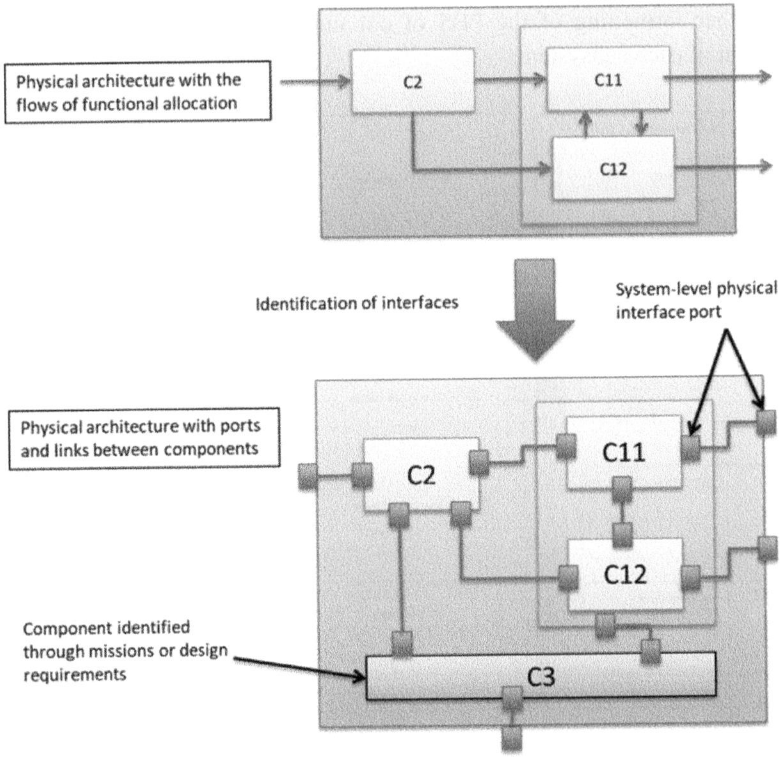

Figure 10.20. *From functional allocation to physical architecture*

Among the interface ports, we have the system-level ports that must be aligned with the subsystem-level ports.

10.3.6. *Application to the unifying thread example: components and interfaces*

Figure 10.21 represents a three-system-level abstraction of the renewable energy water pump functional flows from the functional architecture. It is a static model of the physical architecture on which we can guess the different functional chains for the realization of the missions of the system.

We have, for example:

– A chain that consists of the water table, the centrifugal pump and the water tower. This chain is supplied with energy:

- either through the weather, the wind turbine and the control box,

- or through the electrical network and the control box.

– Another chain that consists of the weather, the wind turbine, the control box and the electrical network.

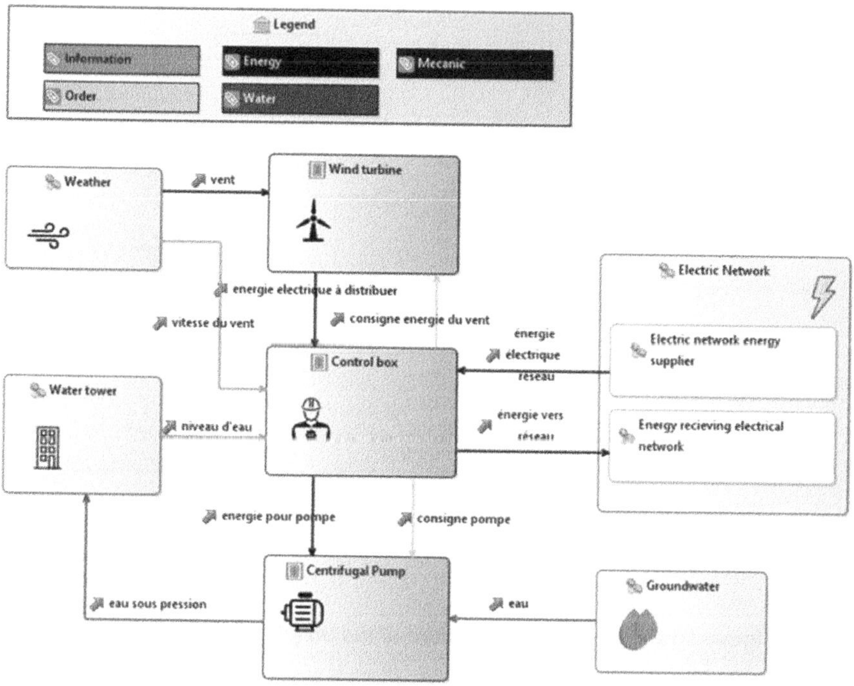

Figure 10.21. *Subsystem-level abstraction of the functional architecture flows*

Figure 10.22 shows the physical architecture of the three subsystems of the renewable energy water pump system. We have identified the different types of links with the interface ports of each subsystem. Among the interface ports, we have identified the external interface ports that interact with the environment and therefore at the system level. On this modeling of the static architecture with ports and links, we have identified in addition a mechanical interface between the wind turbine and the ground.

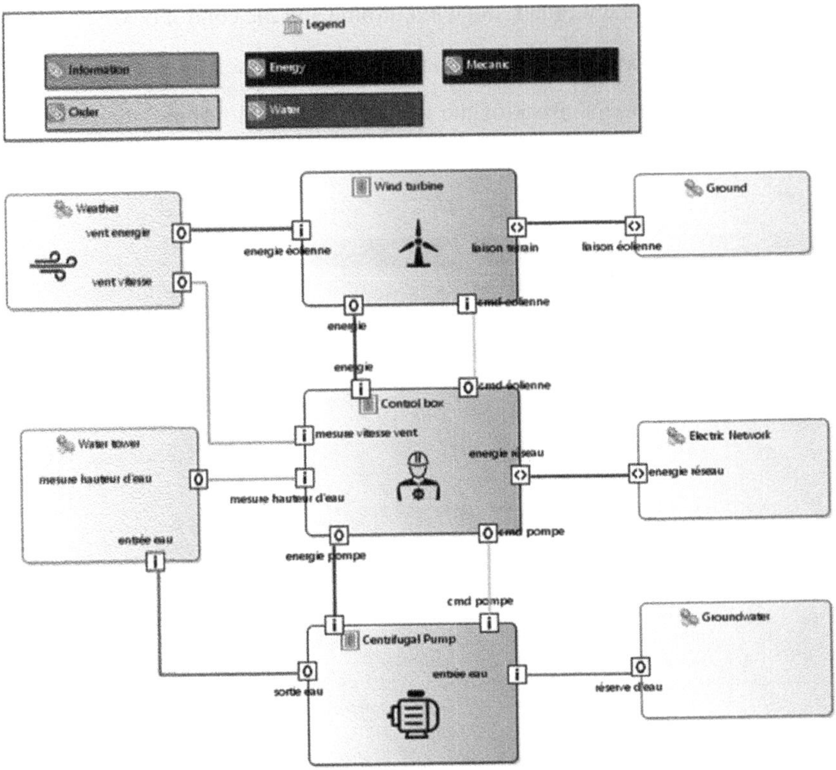

Figure 10.22. *Physical architecture of the unifying thread example*

In Figure 10.23, we have detailed the wind turbine subsystem with its wind generator and its mast. At the level of the wind turbine, we find the four interfaces, namely two energies (wind and electric), one for the control and the final one which is mechanical. This mechanical connection within the subsystem is found on the mast. Indeed, it is the mast that makes the mechanical link between the wind turbine and the ground. The wind generator has been broken down into a propeller, a clutch and an electrical generator. All components have been made independent by identifying all interfaces.

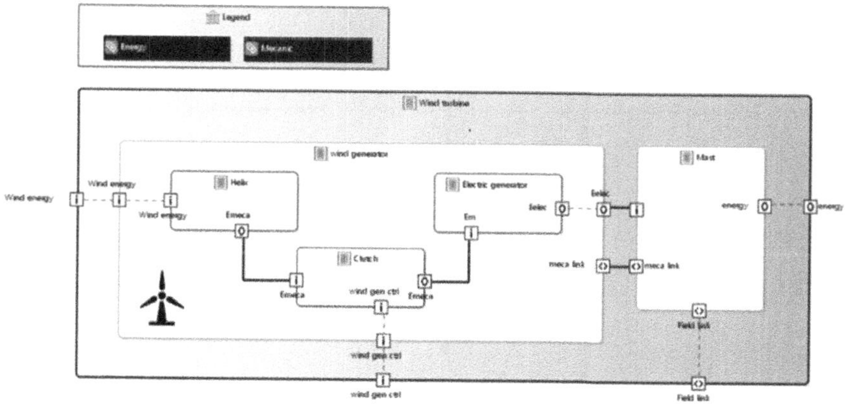

Figure 10.23. *Internal architecture of the unifying thread example subsystem*

10.4. Characterization of interfaces

Functional decomposition allows us to identify sub-functions that perform the system functions which themselves implement the missions. The sub-functions interact with each other, and each interaction creates functional interfaces. These functional interfaces are material, energy and information flows and must be characterized. The physical decomposition reveals hardware, computer and human components that interact with each other. The physical interfaces are derived from the allocation of the functional architecture over the physical decomposition. Functions that exchange MEI flows and are allocated to different components result in physical interfaces (see Figure 10.24). These physical interfaces may require additional functions to adapt the exchanges between the components. These adaptation functions are called interface functions, and they perform the transformations on the flows that are necessary for the exchanges between the components.

To pass these possibly adapted flows through the interface functions, we need to identify and characterize the physical vectors. These physical vectors are called interface links that perform the transport of flows (see Figure 10.25).

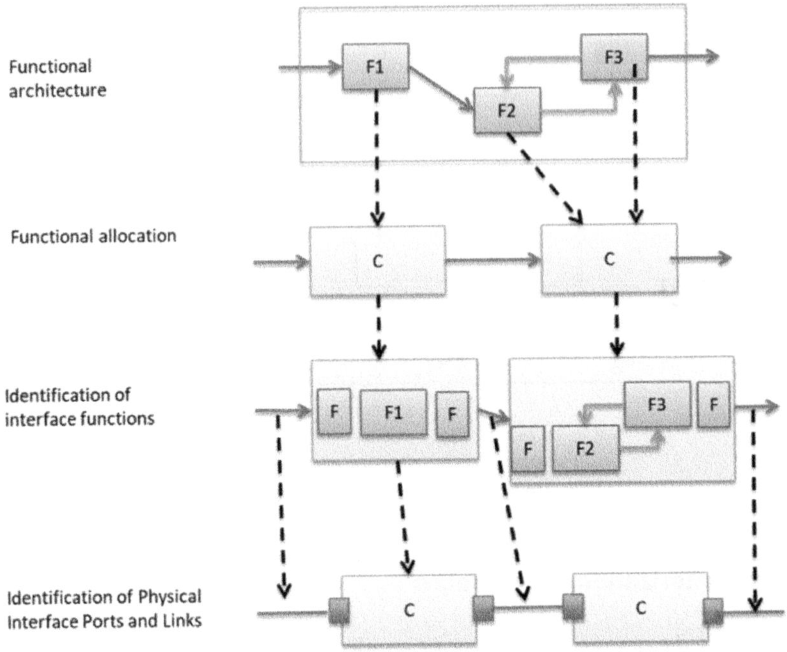

Figure 10.24. *The physical interfaces*

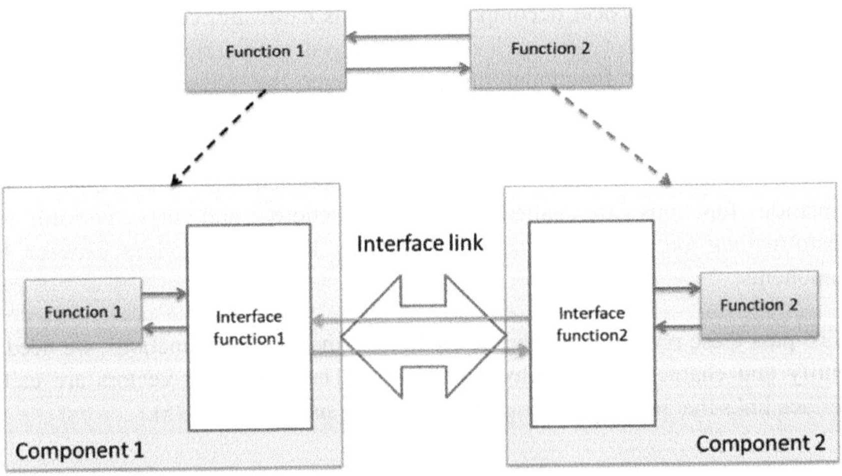

Figure 10.25. *Decomposition of a physical interface*

An interface link can:

– transport multiple streams;

– be a physical component.

The table in Figure 10.26, taken from Faisandier (2013), presents an initial typology of interfaces according to the nature of the components, i.e. hardware, computer or human.

Component A \ Component B	Material	Computer science	Human
Material	Physical or technological link	Sensor	Passive sensor (screen, sound, indicator light, etc.)
Computer science	Actuator, control	Message, protocol, API, memory address	Use a passive hardware sensor (screen...)
Human	Active sensor	Uses an active hardware sensor (keyboard, mouse, etc.)	Sensor with common protocol or semantics

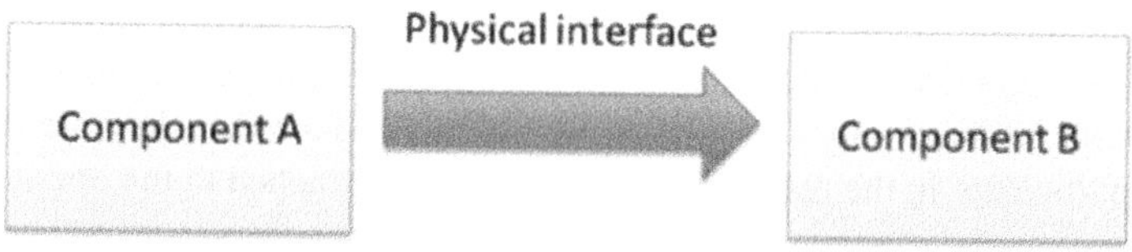

Figure 10.26. *Type of interface according to the nature of the components (Faisandier 2013)*

For the interfaces between two hardware components, the interface link will depend on the technology used, for example, mechanical, electrical, pneumatic and hydraulic links. In order to finalize the physical interface, we identify physical interface ports at the component level that are connected to the physical links. These physical interface ports allow the components to be independent. Physical interfaces must be specified. The table in Figure 10.27 is an example of a physical interface specification with key features to be defined (Krob 2017).

Characteristic	Definition of the key characteristics of a physical interface
Nature	Nature of the exchange supported by the interface
Physical interactions	Flows exchanged by the interface link
Physical functioning	Interface Operation
Configuration	Description of the states that drive the exchange mechanism
Physical constraints	Non-functional requirements of the interface prescribing how the interface must be realized

Figure 10.27. *Specification of a physical interface (Krob 2017)*

10.5. Identification of system configurations

10.5.1. *Definition of a technical configuration*

A system configuration is a constructional state during which the system solicits components to perform one or more functional modes through their activated functions. These functional modes cover one or more operational contexts of the system during which we find the missions to be executed. By transitivity, this means that the technical configurations of the system implemented cover the operational contexts. Still by transitivity, it is possible to deduce that they are the components that participate in the realization of the missions attached to the operational contexts.

In Figure 10.28, we have on the top right the network of functions of the system that performs the missions. Underneath, we have the network of internal technical functions that realizes the functions of the system that execute the missions. And to finish the architecture, we have the network of components that implements the technical functions of the system that performs the missions. At the top left of Figure 10.28, we have the system's life cycle with the operational contexts in which the system's missions are carried out through its functions. Underneath, we have the system's functional modes that cover the operational contexts. Then, we have the technical configurations of the system that implement the functional modes that cover the operational contexts.

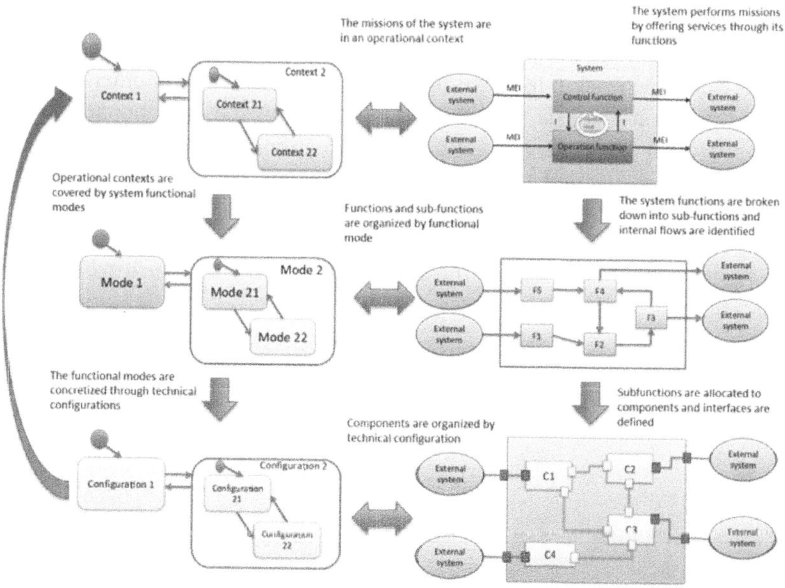

Figure 10.28. *Context, mode and configuration*

10.5.2. *Activities to identify technical configurations*

As an input to this activity, we have:

– operational contexts and missions of the system;

– the functional modes and sub-functions of the system;

– the physical decomposition of the system.

At the end of this activity, we must have:

– the technical configurations of the hierarchical system;

– the components activated in each configuration;

– events that change the configuration of the system.

As for the functional modes, we can apply the GEMMA method which states that we have on our system configurations with and without energy (Duysinx 2016).

For the configurations with energy, we can analyze all the operating modes and list the components used. Modes that use the same components are part of the same configuration. In a second analysis, it is possible to look at the state of each component to better characterize the configurations. Next, we need to identify the events that cause the system to move from one configuration to another. The results of this configuration analysis are cross-referenced with the operational contexts in order to verify that there are no missing configurations and that all configurations cover at least one operational context.

10.5.3. *Application to the unifying thread example: identification of configurations*

In Figure 10.29, we have four configurations that have been identified for our unifying thread example system.

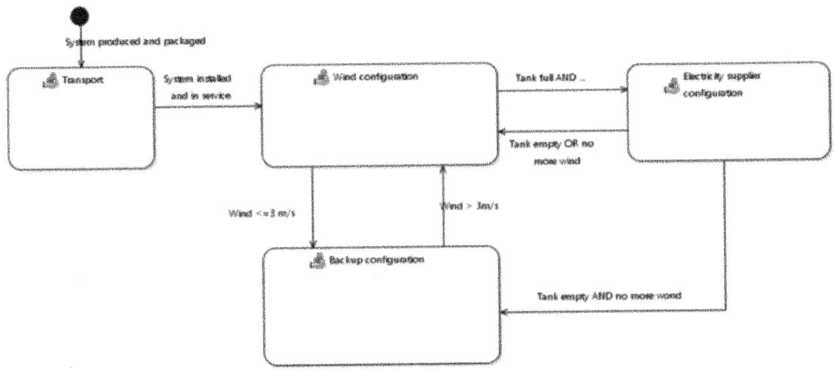

Figure 10.29. *Unifying thread example configurations*

When the system is produced and packed, we have the transport configuration where the system is disassembled in its handling boxes. This is a particular technical configuration where the system is without energy. Then when the system is installed and commissioned, we have the standard wind turbine configuration that will allow the tank to be filled.

In Figure 10.30, the architecture of the standard configuration is shown with all components activated. All of the components of the wind turbine are used to produce electricity that is sent through the control computer to power the centrifugal pump that will fill the water tower with groundwater.

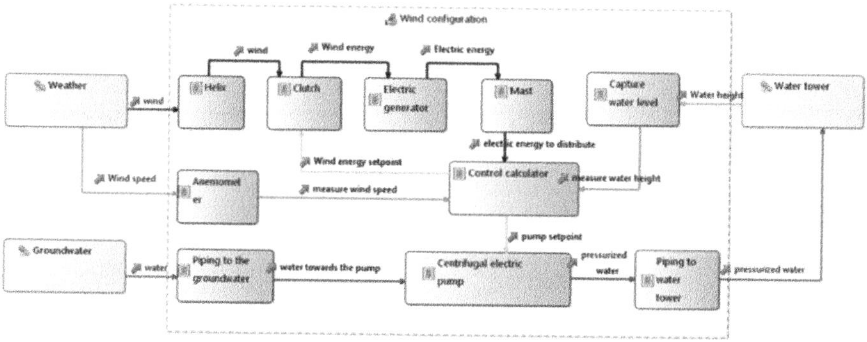

Figure 10.30. *"Wind turbine" standard configuration*

Then, we have the configuration where the wind turbine provides electrical energy to the grid. This configuration fulfills the secondary mission of the system.

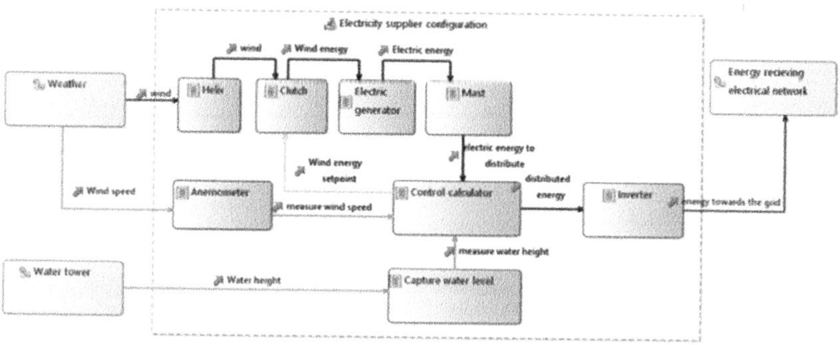

Figure 10.31. *"Supply electricity to the network" configuration*

Figure 10.31 represents the physical architecture of this configuration which has the only difference of conveying the electricity produced by the wind turbine to an inverter in order to inject it on the electrical network.

Finally, we have the emergency configuration where the wind turbine is stopped and the electrical network provides the energy to fill the tank.

Figure 10.32 shows the architecture of this backup configuration where we see the electrical network taking over so that the centrifugal pump can continue to fill the tank with water from the water table.

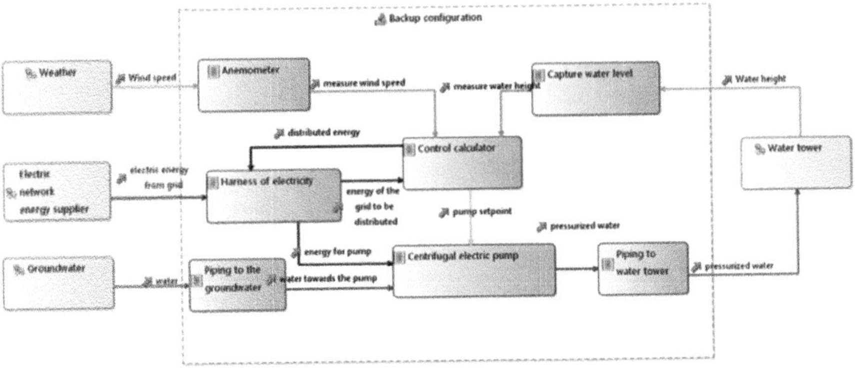

Figure 10.32. *Backup configuration*

10.6. Identifying the physical functioning of the system

10.6.1. *Definition of the physical functioning of the system*

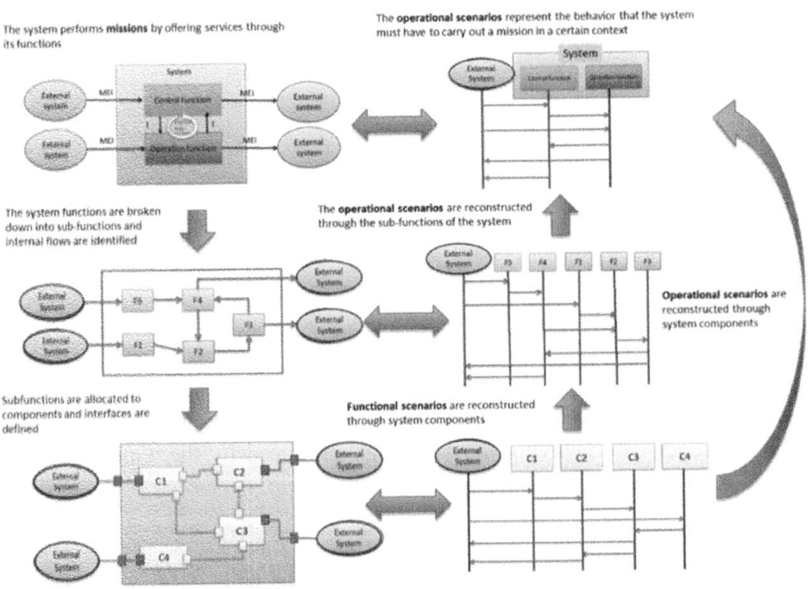

Figure 10.33. *System functioning*

The physical functioning of the system allows for analyzing the behavior of the system through its components to realize the functional and operational scenarios (see Figure 10.33). This analysis enables us to:

– model the behavior of each subsystem in a given configuration through their components; this step verifies the correct decomposition of the system and the interactions between the components;

– reconstruct the functional scenarios in order to verify the correct execution of the functions and the associated functional behaviors through the components;

– reconstruct operational scenarios to verify proper execution of the system's missions.

10.6.2. *Application to the unifying thread example: constructional scenario*

Figure 10.34 represents a physical scenario of our unifying thread example system. This scenario reflects the system behavior for distribution-ready power generation for:

– powering the pump;

– filling the tank;

– selling to the electrical network in order to reduce the citizens' water bill.

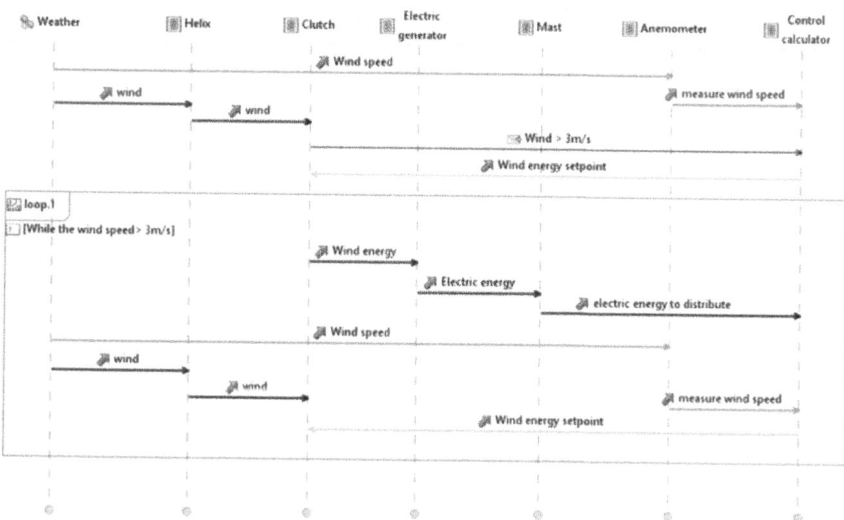

Figure 10.34. *Physical operation of the system*

In this scenario, as long as the wind speed is higher than 3 m/s, the system produces electrical energy from the wind energy.

10.7. Allocation and definition of subsystem requirements

10.7.1. *Emergence of properties*

The system is composed of a set of elements that interact with each other and with the environment. The properties of the system come from its **components** and the **network of interactions** that we call the **structure** of the system.

By taking up the classification of Thomé (1993), we can see in Figure 10.35 the different types of properties at the level of the system and their origins, i.e. from the components of the system or the network of interactions between the components or both (Pollet et al. 2016).

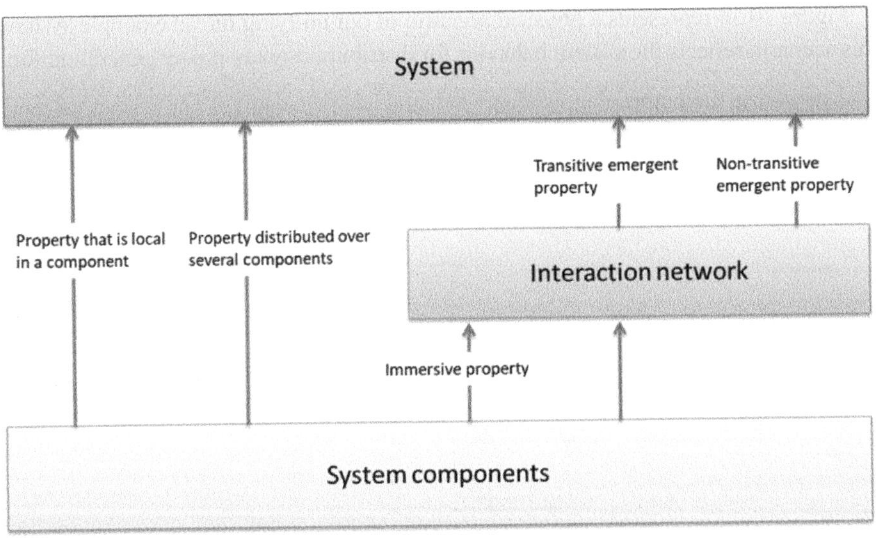

Figure 10.35. *Classification of the properties of a system*

We have the **properties** that are **located on a component**. That is, one of the system's functions can be taken over by one of its subsystems. For example, for a tank, the firing function is taken care of by the gun or the protection function by the armor of the mechanical structure.

We have **properties** that are **distributed over a set of components**. For example, the mass of a system is the sum of the masses of its hardware components.

We have the **immersive properties**, which are component properties that are annihilated by the interaction network. For example, the vibrations of an internal combustion engine can be reduced by the "silent block" connected to the vehicle chassis.

We have the **transitive emergent properties** that arise from the components **and** the interaction network. For example, the dimension of a system is an emergent property because it depends on the dimension of each component and how they are structured together.

We have the **non-transitive emergent properties** that arise only from the interaction network. For example, for a hair dryer, the "blow hot air" function is an emergent function that results only from the flow exchange network. Indeed, this system function is mainly the result of the decomposition into three sub-functions which are "convert electrical energy into mechanical energy", "convert mechanical energy into pressure energy" and "convert electrical energy into thermal energy". It is the interaction network of these three sub-functions that makes it possible to have the "blow hot air" function.

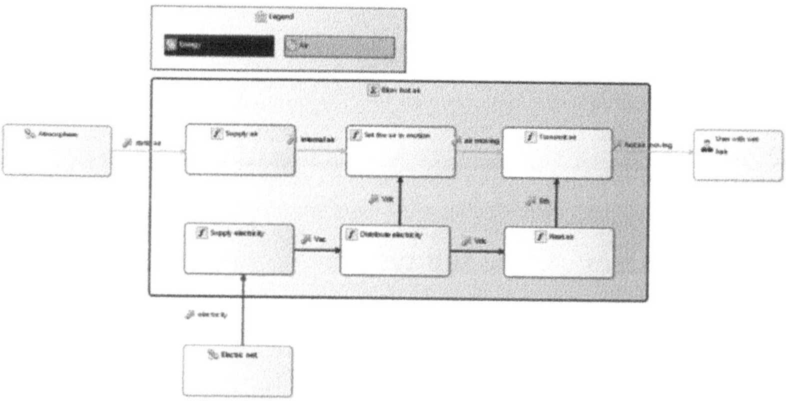

Figure 10.36. *"Blow hot air" function of a hair dryer*

More precisely, Figure 10.36 shows the complete functional decomposition of the "blow hot air" system function of a hair dryer.

Figure 10.37 shows the simplified physical architecture of a hair dryer with the allocation of the sub-functions of the "blow hot air" system function. We can see that this function is indeed non-transitive emergent because none of the physical components of the system has this functional property. We can clearly see on the physical model how this function emerges from the interaction network.

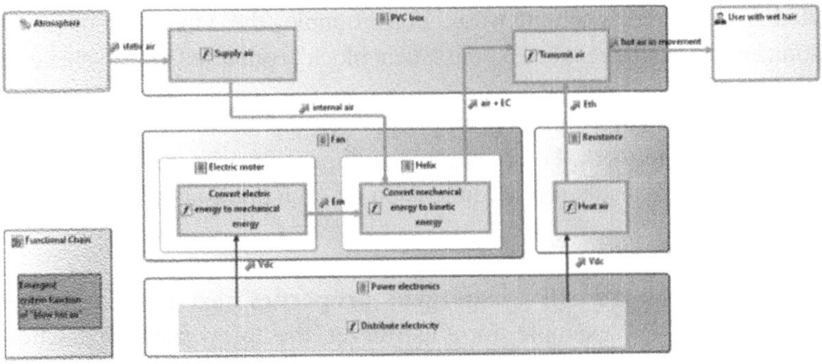

Figure 10.37. *Simplified physical architecture of a hair dryer with the sub-functions*

10.7.2. *Architecture sizing*

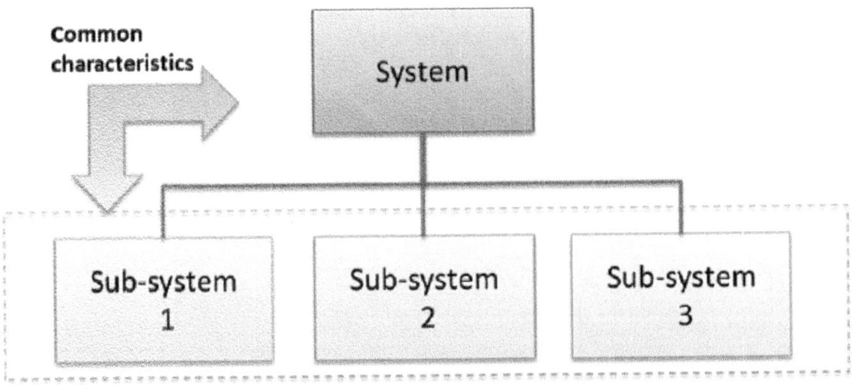

Characteristic	Nature of sizing	Integration law
Characteristic A	Minimization or Maximization	A of system= function (A of sub-systems)
Ex : Weight	Minimization	Weight(syst)=SUM(Weight(Sub-systems))
...	...	...

Figure 10.38. *Sizing*

The dimensioning of a system is an optimization activity under constraints. The objective is to select and implement a solution that is optimally dimensioned with respect to all needs and requirements (see Figure 10.38).

First, we need to identify key features to size on the system and subsystems. Then, we will try to associate optimal values to these characteristics for each component while checking the feasibility. To do this, we will identify for each characteristic the nature of the dimensioning to be carried out, i.e. **maximizing** or **minimizing** the characteristic. Next, we will identify the integration laws that relate the characteristics of a system to the values of the characteristics of the subsystems that compose it. Once the sizing problem is posed, we can solve it as an optimization problem.

10.7.3. *Allocating non-functional requirements*

The specification of the system has allowed us to list a set of properties among which there are non-functional requirements, such as design properties which are, for example, mass, dimensions and color. The design activities led us to identify subsystems that are functional groupings and that respond to constraints imposed by the environment and the other subsystems. The constraints of the environment have been transformed into non-functional properties of the system, and the constraints of the other subsystems have become non-functional requirements induced by the design choices.

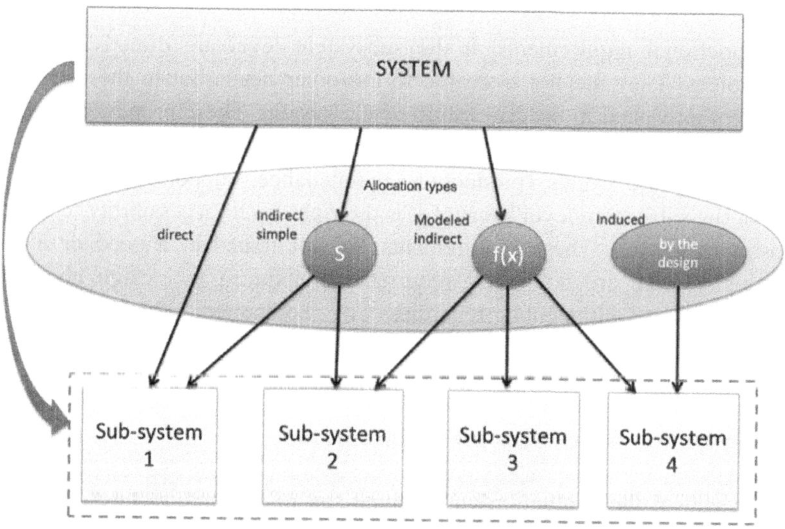

Figure 10.39. *Types of allocation of non-functional requirements (Faisandier 2013)*

The non-functional requirements of the system are allocated to the subsystems in a direct or indirect way (see Figure 10.39). We have four cases of design property allocation.

Case 1: Direct allocation

Non-functional system-level requirements can be directly allocated without transformation to one or more subsystems. For example, the color of the car system is directly allocated to the body subsystem.

Case 2: Simple indirect allocation

Non-functional system level requirements can be distributed to the subsystems through a simple integration law such as a sum or product. For example, the mass of the system is distributed at the level of the material subsystems by a simple law of summation.

Case 3: Modeled indirect allocation

Non-functional requirements at system level can be allocated at subsystem level from mathematical modeling based on business models or sizing calculations, for example.

Case 4: Allocation induced by design choices

Non-functional requirements at the subsystem level are directly induced by design choices. They are not traceable to customer needs but to the needs of the stakeholders interested in the realization of the system. These properties can be the result of business constraints or opportunities. In practice, the stakeholders of the non-operational life cycles (production, maintenance, support, logistics) specify directly at the subsystem level because it is the right level (accessibility, cleanliness, demonstrability, etc.). These requirements are not traced to a need or a system requirement. These are not really requirements induced by design choices, but earlier generic requirements, regardless of component design (cleanliness, compatibility, material, etc.).

10.7.4. *Definition of subsystem requirements*

At the end of the evaluation and comparison of the architectures, a candidate solution was selected to be implemented and integrated (Figure 10.40). This candidate architecture finalizes the organization of the project through the choices of "do" or "buy". In the case where the choice is to buy the subsystem, the set of requirements allocated by the system builder can be completed by state of the art requirements to form the initial requirements of the specification of needs. The supplier of the subsystem responds to this specification with a specification of needs. This specification, which contracts between the system builder and the subsystem, constitutes the specified requirements which can be traced to the system-level requirements.

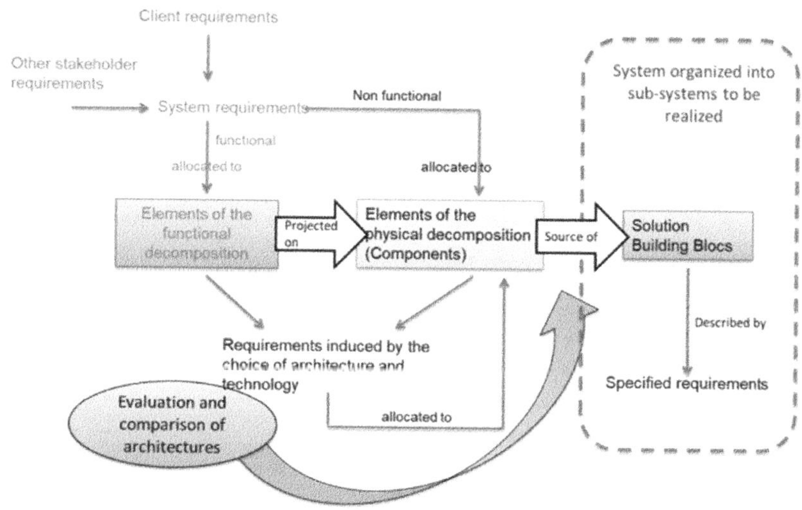

Figure 10.40. *Specified subsystems*

In the case where the choice is to make the subsystem, the procedure is almost the same. The writing of the specified requirements of the subsystems is internal.

The systems engineering process is repeatable on each systemic level, as shown in Figure 10.41. Indeed, at the subsystem level, a specification will be issued in response to the initial requirements, which are the allocated requirements. In the case where the choice is "to buy", the subsystem takes up the needs analysis by modeling an operational architecture of the subsystem to be realized. Then, from this specification, the subsystem performs a design that allows it to allocate its requirements to the components of its subsystem.

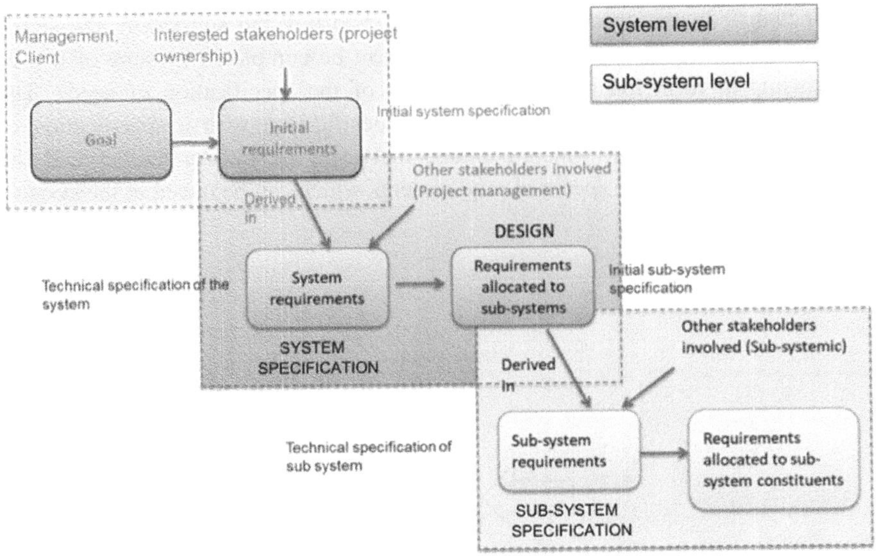

Figure 10.41. *Process applicable to each systemic level*

10.8. Conclusion on the physical design

The physical design is a synthesis process that takes into account the results of the operational and functional architecture as well as the technical possibilities of the company or the state of the art. This systems engineering activity is broken down into several stages. First, we identify the components that allow the technical functions of the system to be realized. Then, we build the interaction network of the components and their hierarchies. We obtain the constructional division of the system that we call the PBS. We identify and characterize the physical interfaces of the system. From the functional modes and system components, we can identify the technical configurations of the system. We reconstruct the functional and operational behavior of the system through its components. Finally, we decline the properties of the system on the subsystems following Thomé's classification. This declination is made in the form of requirements to be followed by the project management. During component identification, several solution choices can be identified. These different solutions will give several candidate architectures that will have to be evaluated and compared.

11

Verification and Validation of the Physical Architecture

11.1. Verification of the physical architecture

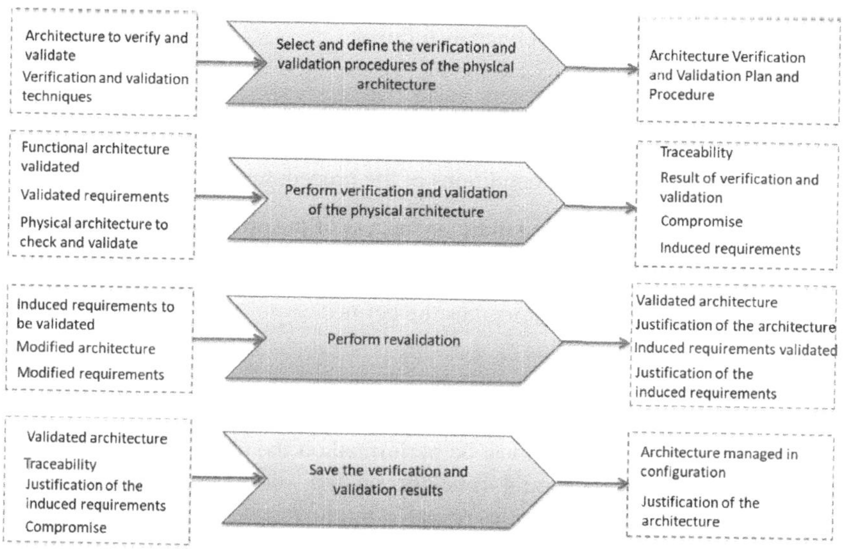

Figure 11.1. *Verification and validation of the physical architecture. For a color version of this figure, see www.iste.co.uk/hygounenc/system.zip*

The objective of the physical architecture verification is to look at:

– consistency with the functional architecture;

– allocation of non-functional requirements;

– compliance of the physical architecture with the life cycle processes and contributing products (V&V readiness, deployability, maintainability, recyclability).

Figure 11.1 presents a physical design verification process inspired by the EIA-632:1999 standard (ANSI 1999).

11.1.1. *Defining physical design verification procedures*

We need to select and define procedures for verifying the physical architecture using the following methods:

– **inspection**, e.g. of technical drawings and diagrams;

– **analysis**, e.g. using simulation or virtual prototyping;

– **demonstration**, e.g. using physical models;

– **testing**, e.g. using physical prototypes.

A verification procedure is defined as follows:

– a selected method;

– a verification objective;

– the preconditions and postconditions of the procedure;

– criteria for determining the failure or success of the procedure.

To carry out the verification procedures, a verification environment such as computer equipment and simulators must be defined.

11.1.2. *Performing physical design verification*

The verification procedures must be performed on the physical architecture. The results of the verification should conclude on the verification objectives. Gaps, conflicts, discrepancies or anomalies revealed by the results of the verification must be collected and evaluated. In particular, the verification should provide the following evidence:

– physical design elements whose internal and external interfaces are traceable to the selected architecture requirements;

– non-functional system requirements are traceable to the design elements;

– the physical architecture implements the functions of the functional architecture;

– the physical architecture meets the functional and performance requirements;

– the physical architecture meets the various constraints.

If any requirements are not covered by the selected physical architecture then the omissions must be corrected. If elements of the architecture are not traceable to the functional architecture or the system requirements, then either derived requirements are introduced or corrections are proposed to the functional architecture or the system requirements repository.

11.1.3. *Rechecking the physical design*

When variations and anomalies in the results of verification procedures are attributed to poor verification conduct or an inadequate test environment, the verification procedures are repeated to obtain valid results.

11.1.4. *Recording the results of the physical design verification*

At the end of the physical architecture verification and validation activity, the following information can be recorded in the project database:

– verification procedures;

– the verification results that were obtained;

– corrective actions;

– the justifications of the physical architecture;

– trade-off analyses;

– the definition of the physical architecture verified.

11.2. Validation of the physical architecture

The validation activity of the physical architecture is done at the same time as the verification. The objective of this activity is to validate the conformity of the physical architecture characterized by its requirements with the system requirements. We must analyze whether the system requirements allocated and broken down into subsystem level requirements on the selected physical architecture elements are properly covered. And we must analyze whether the subsystem requirements that define the selected physical architecture cover the system requirements.

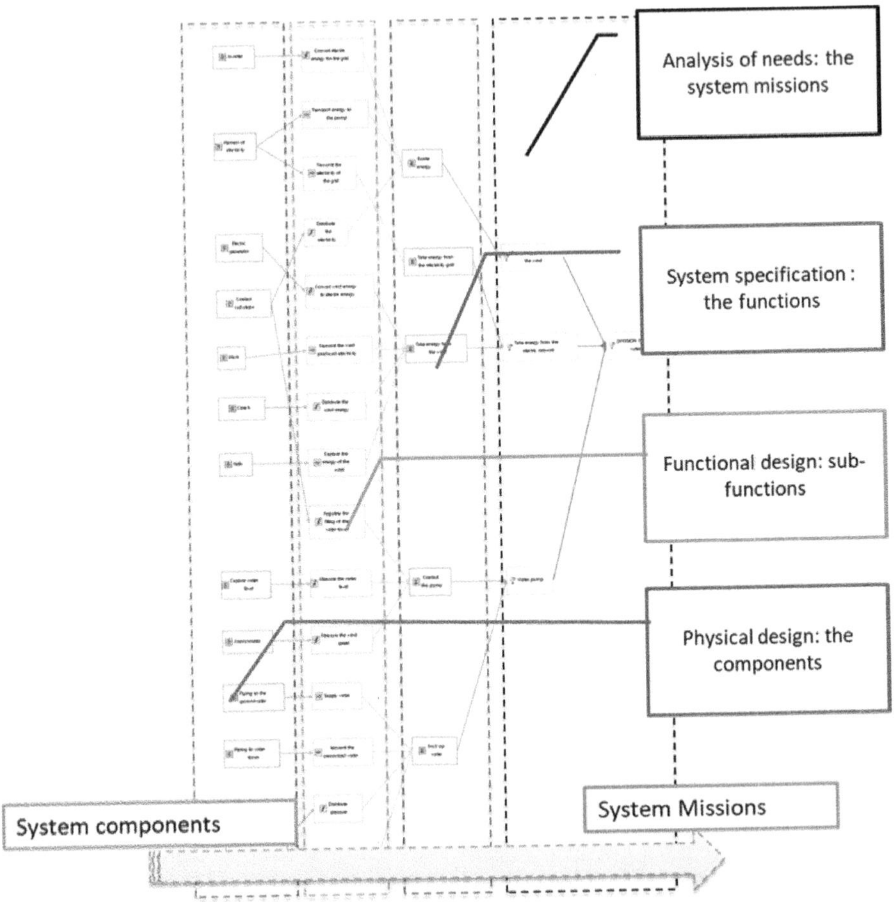

Figure 11.2. *From the solution to the mission of the unifying thread example. For a color version of this figure, see www.iste.co.uk/hygounenc/system.zip*

Figure 11.2 shows a tree structure relating to our unifying thread example that allows us to make the link between the components of the physical architecture, the sub-functions of the functional architecture, the functions of the system specification and the missions of the requirements analysis. This type of simple representation validates at least that the missions of the system are well realized by the components of the system.

11.3. Conclusion on the verification and validation of the physical architecture

Each systems engineering activity must be verified and validated. For the physical design, we will verify that the candidate architecture implements the technical functions of the functional architecture. We will also validate that the requirements stated on the candidate architecture meet the system requirements.

12

Evaluation and Comparison of Solutions

12.1. Evaluation of architectures

12.1.1. *Selection of common criteria*

Several functional decompositions can meet the same needs, and several physical components can implement the same technical functions and meet the same non-functional requirements. This gives us several functional and constructional architectures that we must compare and evaluate. To evaluate multiple architectures, we will need to identify a list of criteria, common to all architectures. We have two types of criteria, namely:

– value criteria;

– risk criteria which incorporate cost criteria.

Each of these criteria will need to be evaluated at the level of each candidate architecture. The selected scoring should be discriminating so that the same average is not used for each solution.

12.1.2. *Value criteria*

Value criteria are criteria for which the architectural solution provides value to the client or the project.

For a color version of all the figures in this chapter, see www.iste.co.uk/hygounenc/system.zip.

Systems Engineering,

Contribution to VALUE	1pt	3pts	9pts
Improved services for stakeholders	None	Average	Significant
Improved internal technical performance	None	Average	Significant
Reuse, standardization, pooling	None	Average	Significant
Simplification of the architecture	None	Average	High
Decrease in time to market	None	Average	High

Figure 12.1. *Example of value criteria (Krob 2017)*

The table in Figure 12.1 shows the proposed value criteria from Krob (2017). We can also see a discriminating notation. This list can be used for brainstorming as a classification of criteria.

12.1.3. *Risk and cost criteria*

Risk criteria are criteria for which the architectural solution presents a risk with respect to the client's need or the project's constraints. Regarding the cost criteria, there can be several, such as:

– non-recurring costs;

– recurring costs;

– maintenance costs.

Risk and cost	1pt	3pts	9pts
Cost	Low	Average	High
Increased internal couplings	Low	Average	High
Complexity of realization	Low	Average	High
Maturity of solutions	High	Average	Low
Time pressure	Low	Average	High

Figure 12.2. *Example of risk criteria (Krob 2017)*

The table in Figure 12.2 shows the proposed risk criteria from Krob (2017). We can also see a discriminating notation. This list can be used for brainstorming as a classification of criteria.

12.2. Comparison of architectures

12.2.1. *Selecting stakeholders*

The comparison of architectures is a collaborative activity. To carry out this project, the system engineer must involve various stakeholders such as the business lines, cross-functional engineers (safety, quality), management, a customer representative (operational architect), marketing, sales, purchasing, etc. The system engineer must select several candidate architectures. Since this type of activity has a significant time cost, it is recommended that a reasonable number of architectures are evaluated and that the least relevant ones are eliminated before the start of the project. These preliminary choices may be justified. There are four steps to facilitate the work:

– evaluate the weight of each criterion;

– get stakeholders to vote;

– compare the architectures;

– justify the choice of architectures.

12.2.2. *Characterizing the weight of each criterion*

The first step is to characterize the weight of each criterion:

– **step 1:** identify the stakeholders who will participate in the evaluation;

– **step 2:** identify a set of criteria;

– **step 3:** evaluate the relative importance of the criteria by associating a weight to them.

	Stakeholder 1	Stakeholder 2	...	Criterion weight
Criterion 1	...	...	...	Sum of grades /TOTAL
Criterion 2				
...				
Total of votes	10 points	10 points	...	TOTAL

Figure 12.3. *Evaluation of the weight of the criteria*

The table in Figure 12.3 shows a technique for evaluating the weight of each criterion (Krob 2017).

12.2.3. *Getting stakeholders to vote*

The second step is to present the different architectures and have the different stakeholders vote.

12.2.4. *Comparing architectures*

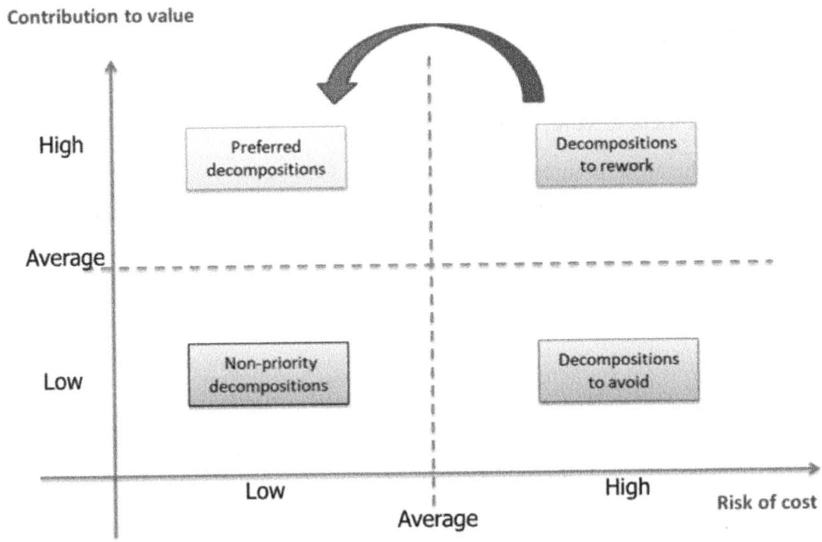

Figure 12.4. *Comparing architectures*

The third step consists of projecting the two scores obtained for each architecture on a decision-making diagram. Figure 12.4 shows the decision-making diagram with the risk and cost score on the x-axis and the value score of each candidate architecture on the y-axis. The diagram is divided into four areas for decision-making.

We have four cases based on the two value and risk scores. An architecture that will have low risk and low value will not be a **priority solution**. A high-risk, low-value architecture will be among the **solutions to be discarded**. A high-risk, high-value architecture is one of the **solutions to be reworked**. In this case, we have two solutions:

– either the solution can be reworked and re-evaluated as part of the project;

– or the solution must be matured through research and development activities.

A low-risk, high-value architecture will be one of the solutions to be selected for the project.

12.2.5. *Justifying architectural choices*

The fourth step consists of closing the project by justifying the choice for each evaluated architecture.

These justifications will be kept in the supporting file that will be produced during the verification and validation activities of the system architecture.

12.3. Conclusion on the evaluation and comparison of the architectures

The process of evaluating and comparing architectures is a high-value systems engineering process. Indeed, it selects in a collaborative way the best architecture or at least the one that presents the best compromise.

This activity is easy to implement; you just have to select a common value and risk criteria for each architecture. The cost criteria are part of the risk criteria. As an option, it is possible to determine a weight for each criterion.

Then, the project manager, along with the system engineer will select a set of stakeholders to participate in the evaluation of the different architectures. Each participant will evaluate the different candidate solutions according to a discriminating score. If the scores are weighted by weights on the criteria, it will be necessary to determine each of them beforehand.

Finally, each architecture, thus evaluated, is compared on a diagram where the risk score is projected on the abscissa and the value score on the ordinate. The selected solution will be the architecture with the lowest risk and highest value.

PART 5

Virtual System Integration

13

Integration on a W-Cycle

13.1. Virtual integration and real integration

In Parts 5 and 6, we will discuss two types of integration which are the virtual integration of the system and the actual integration of the final product. On a W-shaped development cycle, the objective is to minimize the risks on the project by performing several levels of virtual escalation. In Figure 13.1, we have the details of the different levels of escalation of the virtual development cycle with:

– a virtual system-level escalation;

– a virtual subsystem-level escalation;

– a virtual equipment-level escalation;

– a model recalibration loop;

– a real escalation of the integrated final system.

A "virtual cycle escalation" consists of integrating a level of analytical models of the system into a simulated environment and validating operational scenarios. We can consider at least three levels of virtual integration. Each level corresponds to a stage in the project and a systems engineering process.

For a color version of all the figures in this chapter, see www.iste.co.uk/hygounenc/system.zip.

Systems Engineering,
by Emmanuel HYGOUNENC.

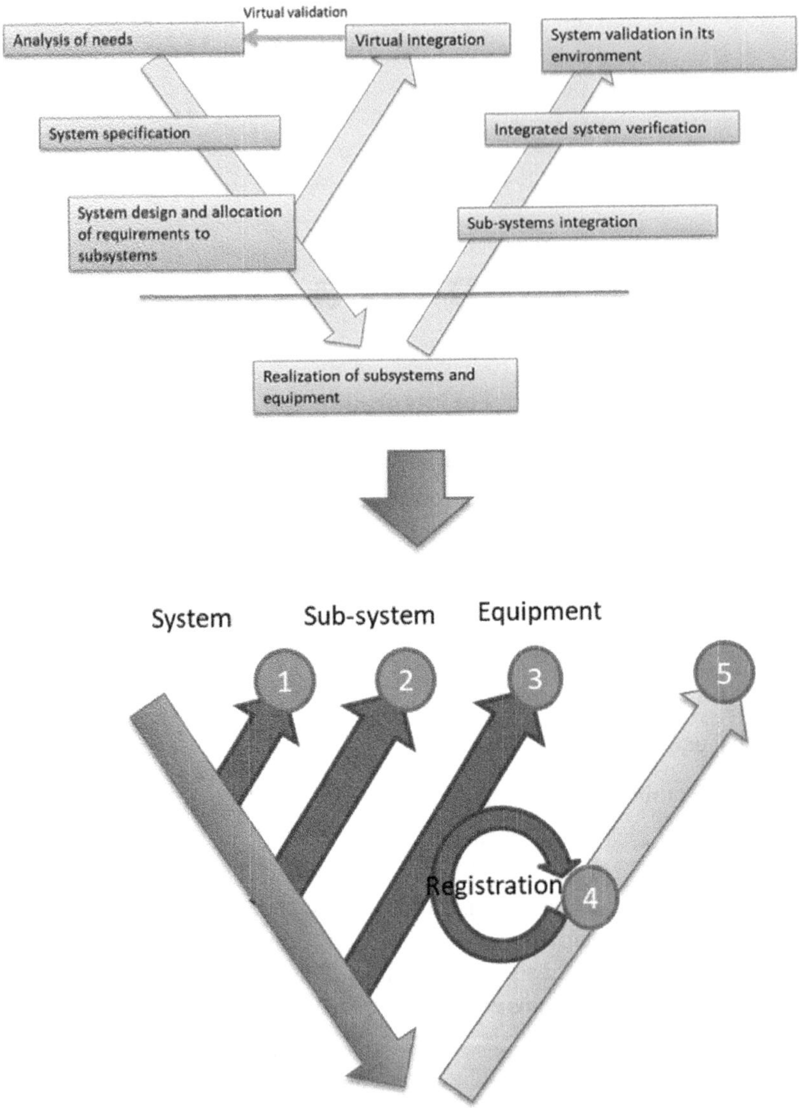

Figure 13.1. *W-cycle and virtual integration levels*

Lift 1 – in Figure 13.1

Level 1 virtual integration is a system-level behavioral integration. It allows the development, verification and validation of the system specification. In Paper (2019), the author talks about running a mirror model. This level of virtual integration corresponds to the process of analyzing the needs and specifying the system.

Lift 2 – in Figure 13.1

Level 2 virtual integration is a subsystem-level behavioral integration. At this level, we test and integrate the functional design of the system. We can verify and validate the requirements of the subsystems. This level of integration corresponds to the functional and physical design process.

Lift 3 – in Figure 13.1

Level 3 virtual integration is an integration of equipment-level behavioral models. At this level, we can test and integrate the functional design of the subsystems and verify and validate the requirements of the equipment. This level of integration corresponds to the detailed design stage. As the equipment of the subsystems has been selected, we can do the integration of multi physical models.

Lift 5 – in Figure 13.1

The actual integration of the system consists of assembling and testing the different subsystems together to form the integrated system. At this level, we can verify that the final system meets all the requirements of the specification, which means that it has all the functional and non-functional properties. This step prepares the system to be validated. Validation consists of looking at whether the system meets operational expectations, i.e. is it able to cover all of its missions with the right performance while respecting the various constraints (bottom cycle in Figure 13.1).

Recalibration 4 – in Figure 13.1

During the actual recovery of the development cycle, the models used for the virtual integration can be rejected. This stage of capitalization of the knowledge of the models makes it possible to know the level of reliability of the different virtual integrations and to refine the implementation of the integration, verification and validation plan. We call this step the recalibration of the models.

13.2. Simulation models

For virtual integration, we will use two types of models that will be connected together, namely a design model and a specification model (see Figure 13.2).

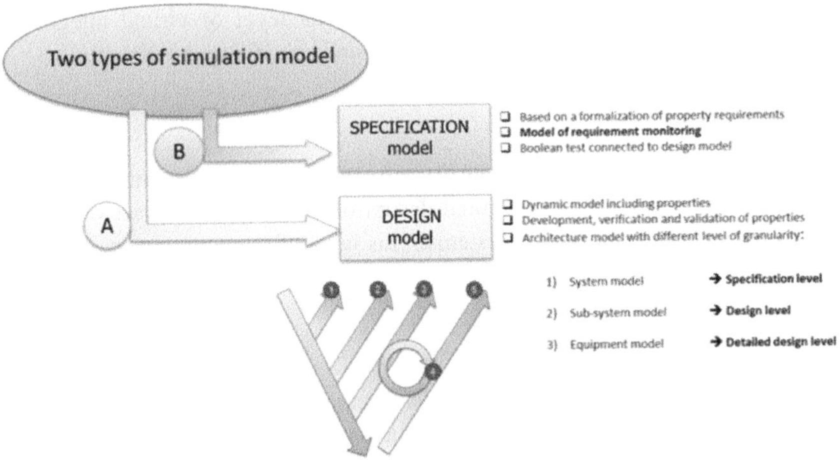

Figure 13.2. *Two types of simulation models*

First of all, we have the **design model** which is a behavioral model of the system. This causal and temporal model simulates the dynamics of the system and its environment. Depending on the degree of maturity, it may be more or less representative of reality. It includes some system properties and allows for focusing or verification or validation. This dynamic model is representative of a certain level of architecture, namely the system, subsystems or equipment. For the same level of architecture, there may be multiple design simulation models.

Then, we have the **specification model**. It is based on a formalization of property requirements. It is a monitoring model, which must be connected to a design model and takes the form of a set of Boolean tests. This lets us know whether or not a property is valid throughout the execution time of the scenario. This also verifies that the property is tested at least once on the scenario. A specification model follows the principles of requirements engineering and can therefore be broken down into a lower-level specification model.

To achieve a virtual integration (see Figure 13.3), the specification model is connected to the design model to monitor the requirements formalized in properties. The design model that carries properties is stimulated at the entrance by operational scenarios. At the exit of the design model, we have the temporal results of the different variables observed. At the exit of the specification model we have over the entire duration of the simulation, the evaluation of the presence or not of each of the simulated properties and, where present, whether it is correct or not.

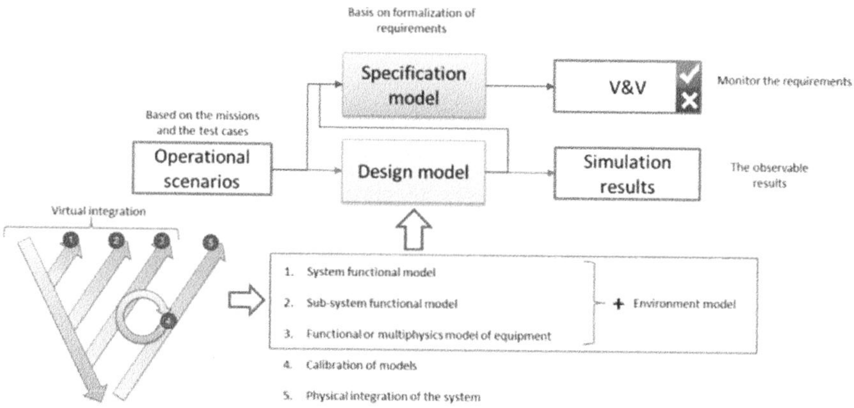

Figure 13.3. *Integration of the specification and design model*

On a design and specification model, it is possible to simulate several operational scenarios. In addition, in the case of a verification, we can replace the operational scenarios with **test cases** that can be verification actions.

DEFINITION.– A test case consists of the input data for a test with expected results from running that test. This set of input data and expected results is associated with a test objective which is the behavior of the system to be tested (Longuet 2016).

13.3. Conclusion on integration on a W-cycle

In this chapter, we have discussed the development cycle in W which has as a particularity the fact of proposing feedback of virtual integrations. Each level of virtual escalation makes it possible to secure the development of the system at different levels of maturity. Virtual integration consists of modeling and simulating the system in its environment. It is based on two types of simulation models, namely a design model and a specification model.

The design model involves the simulation engineer and can simulate the behavior of the system, subsystems or equipment.

The specification model connects to the design model to monitor the properties set by the system engineer. The peculiarity of this model is that it must be developed with a system-level design model. It can be reused identically in all levels of virtual integration.

14

Creating a Simulable Design Model

14.1. Defining the simulation objectives

Defining the objective of the simulation allows us to frame and quantify the goal we are trying to achieve. We have three types of goals for a simulation of the system, namely:

– **development** of specification or design elements;

– **verification** of specification or design elements;

– **validation** of specification or design elements.

14.2. Simulation and the systemic approach

Figure 14.1 takes up the systemic approach proposed by Donnadieu and Karsky (2002) and makes the link with the systems engineering approach presented in the previous chapters. At the top, we have the observable which is the system we seek to design in its environment. Our goal is to validate that this system responds well to the problem posed by user stakeholders. For this, we have the systemic exploration phase which corresponds to the analysis and design of the system. For the formatting of the elements of this systemic exploration, we have the qualitative modeling phase that corresponds to the architectural modeling of our system. We complete this architectural modeling with causal modeling that will allow us to identify the endogenous and exogenous variables that will characterize our system and its environment (Figure 14.3). In addition, these variables will allow us to simulate its behavior. We talk about causal modeling because we define the causal link (positive

For a color version of all the figures in this chapter, see www.iste.co.uk/hygounenc/system.zip.

Systems Engineering,
by Emmanuel HYGOUNENC. © ISTE Ltd 2023.

or negative influence) between the variables. Then, dynamic modeling formalizes causal analysis in order to provide the system of differential equations for simulation. The introduction of time allows us to move on to the execution and analysis of simulation results.

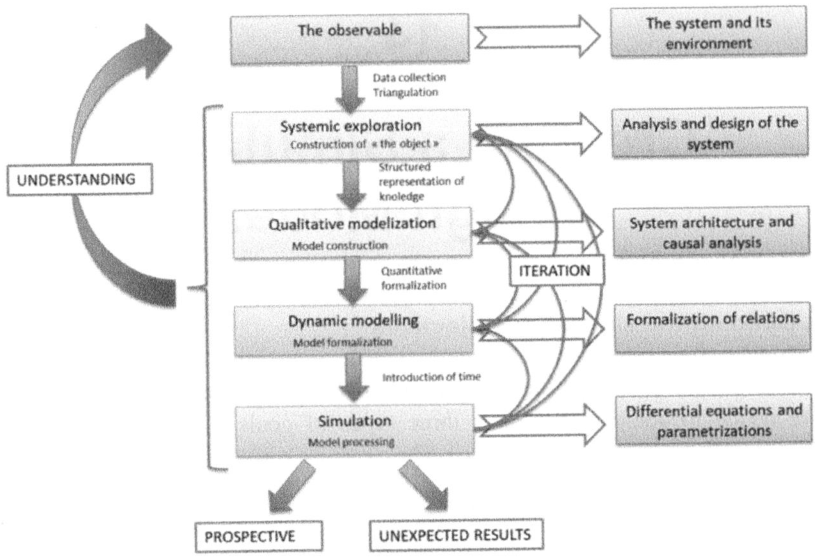

Figure 14.1. *Steps in the systems approach (Donnadieu and Karsky 2002)*

14.3. Analysis and modeling of system architecture

We recall that each architectural view is composed of a set of points of view that are connected to each other. We have:

– the **static** or structural point of view of the system;

– the **dynamic event point** of view with system **states** and triggering events;

– the dynamic point of view is the integration of the two previous points of view, namely the **behavior** of the system in a given state.

Architecture modeling is a necessary entry point before formalizing the simulation's network of equations. This modeling allows us to establish:

– all inputs and outputs of the system, function or component;

– the interactions of the system, function or component with its environment;

– the dynamics and expected behavior of the system, function or component.

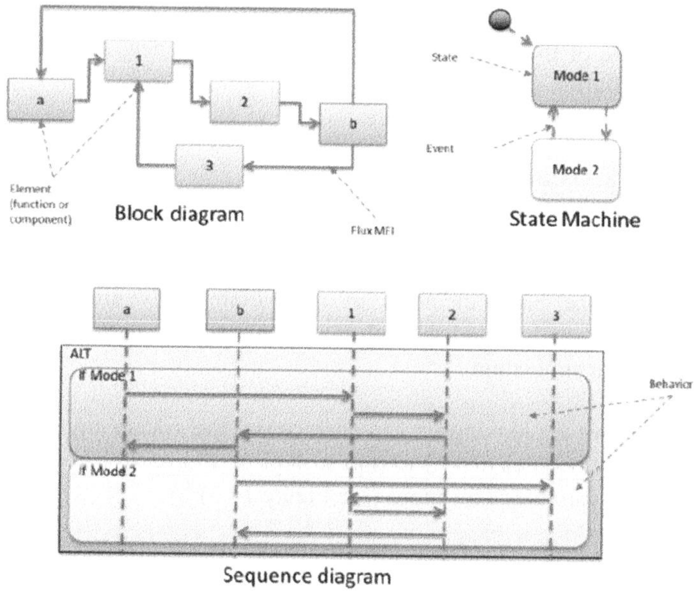

Figure 14.2. *Architecture models*

In Figure 14.2, we have the three main types of qualitative models used to model a system architecture.

The **block diagram** is used to model the system's functional and constructional interaction network.

The **state machine** makes it possible to model the dynamics of the system and to link functional modes and technical configurations to internal and external events.

The **sequence diagram** models the behavior of the system and makes it possible to make the link between the static structural view and the dynamic event view. We can note that this is the only diagram that shows time, in the form of a dotted lifeline.

In systems engineering, there are several languages and formalisms for modeling a system. We will not present them here but invite the interested reader to refer to, for example, Friedenthal et al. (2009) for SysML language or Voirin (2017) for a presentation of a formalism implementing Arcadia methodology.

From the simulation objectives, we can select the vision and points of view necessary to build this same simulation. System architecture models must be complemented by a qualitative model around system and environment variables to meet the objective of the simulation. This gives rise to a new **system perimeter**, which is the **simulation system**.

14.4. Analysis and causal modeling

14.4.1. *Extracting simulation variables*

Variables are the basic elements of simulation. They are identified from architectural models and related through mathematical laws or physics to form all the equations of our simulation.

The variables we are looking for can characterize the system environment as the system itself. The variables that characterize the system are called endogenous and the environmental variables exogenous (see Figure 14.3).

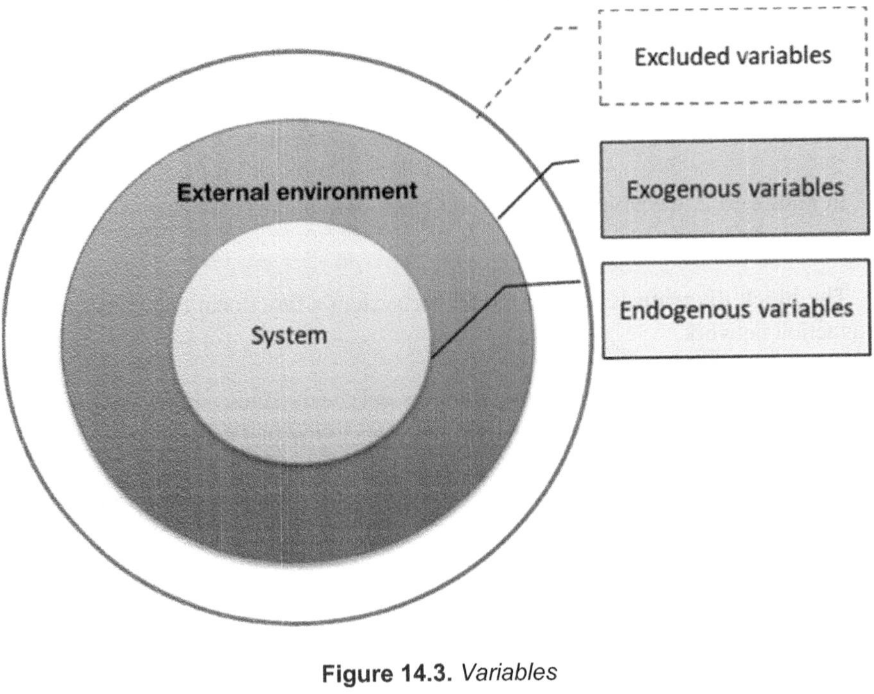

Figure 14.3. *Variables*

14.4.2. *Relationships between variables*

With the list of variables that we have identified from the architecture models, we will look for the relationships that exist between these variables. There are two types of relationship between variables (see Figure 14.4).

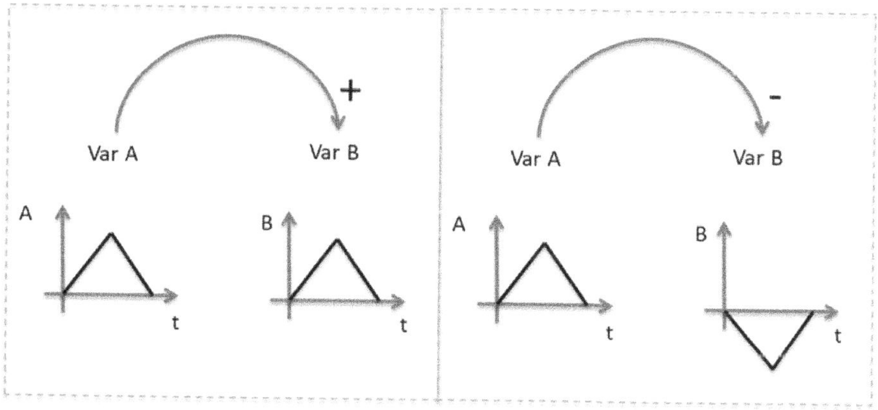

Figure 14.4. *Influences between variables*

We have **positive relationships**, in other words, variable A positively influences variable B. For example, if A is growing, then B is growing, and vice versa if A is decreasing, then B is also decreasing.

We have **negative relationships**, in other words, variable A negatively influences variable B. For example, if A is growing, then B is decreasing, and vice versa if A is decreasing, then B is growing.

14.4.3. *Feedback loops*

Through these relationships, we seek to identify and reveal the feedback loops that are characteristic of **complex** systems and certain **emergent properties**.

We have two types of feedback loops: positive feedback loops and negative feedback loops.

14.4.3.1. *Positive feedback loops*

We have positive loops, i.e. loops with an **even number of negative relationships**. These feedback loops are unstable and are also called explosive or collapse (Figure 14.5).

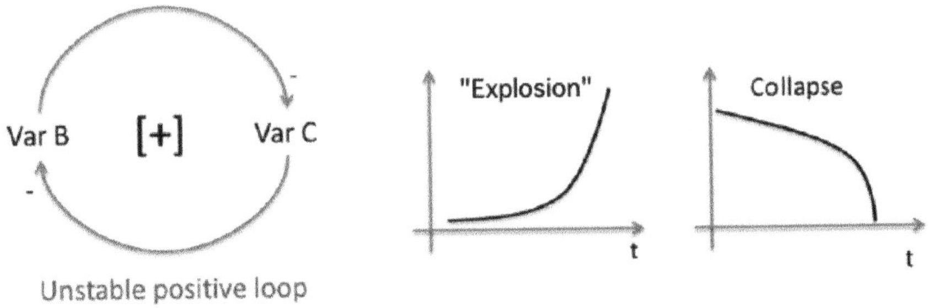

Figure 14.5. *Positive or unstable loop*

14.4.3.2. *Negative feedback loops*

We have **negative loops**. That is, those whose number of negative relationships is odd. These feedback loops are stable and are also called stabilizers or regulators (see Figure 14.6).

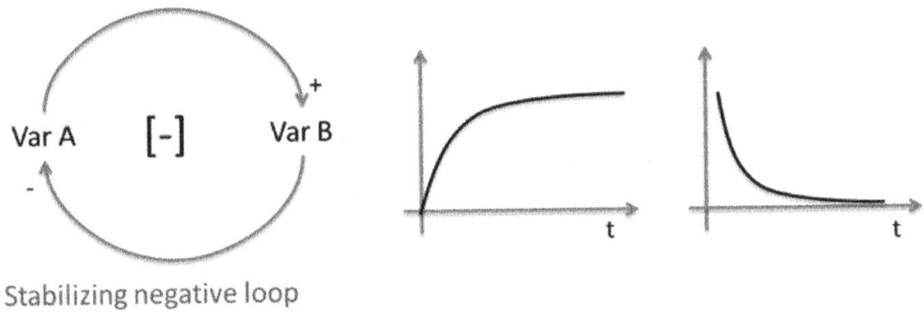

Figure 14.6. *Negative or stable loop*

14.4.3.3. *Interconnected loops*

A system with its environment can have, depending on its complexity, several interconnected loops. For example, Figure 14.7 shows two feedback loops between three variables.

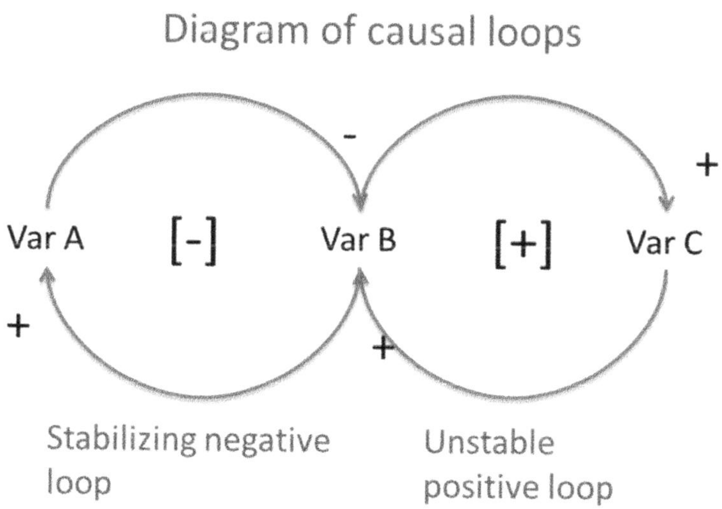

Figure 14.7. *Feedback loops*

14.4.3.4. *Limiting factors and key factors*

In the behavior of a system, certain variables can have a greater influence with greater effects (Garcia 2019).

There are **limiting factors**, which are variables that limit the progression or growth of the system. For example, the saturation of an actuator such as a hydraulic pump that will have a maximum output, even if at the input we increase the power.

We also have **key factors**, which are variables that have a great influence on the system. Thus, a slight modification of these variables has a significant effect on the behavior of the system. For a wind turbine, wind speed is a key factor. A change in this exogenous variable results in a cube-based variation in the electrical power produced.

14.4.3.5. *Delays*

Delays or delay phenomena may exist regarding the relationships between the variables. In Figure 14.8, we have a delay regarding the relationship between variables A and B, which translates into a pure delay in simulation in our example. There are several types of delays; they are defined when formalizing the causal model.

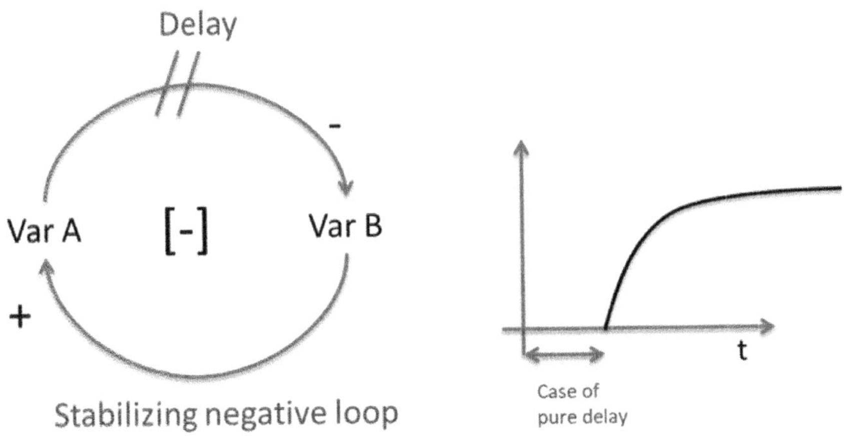

Figure 14.8. *Delay on a relationship*

14.4.4. *Application to the unifying thread example: causal analysis*

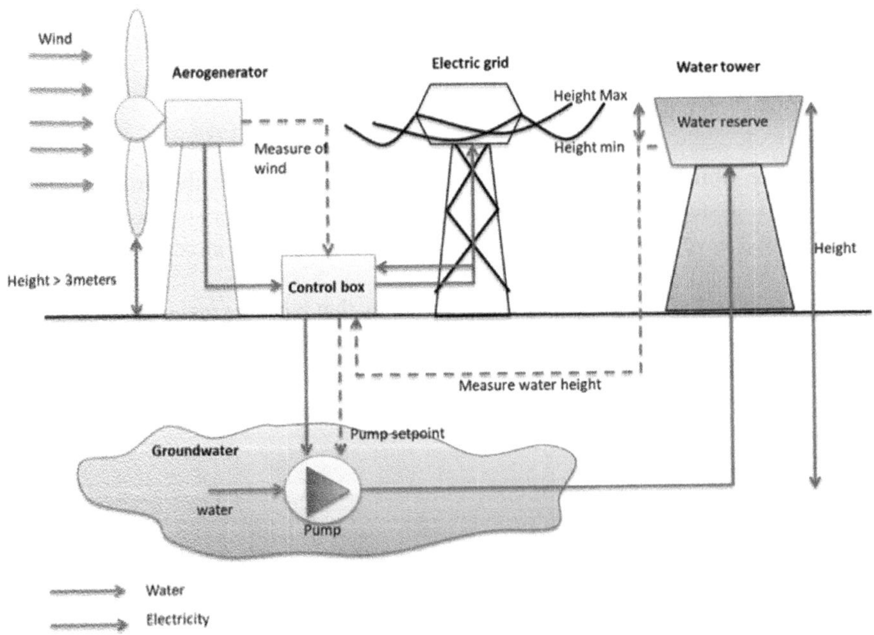

Figure 14.9. *Water pump system with wind turbine*

In Figure 14.9, we have a simplified representation of the water pump system with a wind turbine represented in green in its environment.

We will build our diagram of causal loops, gradually starting with a simple goal which is to understand how the system fills the tower reservoir with wind energy.

We will start from the main mission of the system, which is "fill the reservoir with the energy of the wind" and analyze the variables related to this use case (see Figure 14.10).

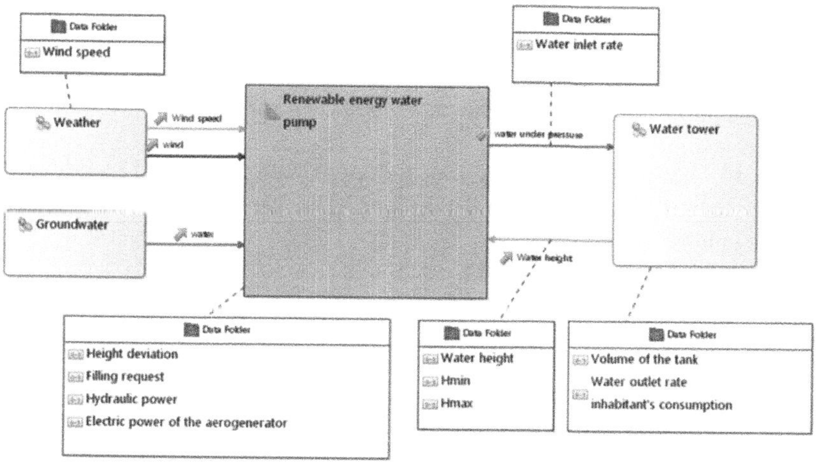

Figure 14.10. *Identification of variables for the mission "fill the water tank" with wind energy*

At the level of the entry state of the mission, we have the tank which has a height of water H < Hmin. At the end of the mission, the tank will have a height H = Hmax and a maximum volume of water.

So we have four variables: ***tank volume, water height, Hmin*** *and* ***Hmax*** *(see Figure 14.10).*

The request to fill the tank is made when the height is less than Hmin and stops when the height is equal to Hmax, i.e. when the difference between the desired maximum height and the actual height is zero.

*We have identified two more variables: the **filling demand** and **the height** difference between the maximum height and the water height (see Figure 14.10).*

Hydraulic power is provided by the electrical power of the wind turbine, which comes from the wind speed. The hydraulic power is transformed into the flow of water at the inlet of the tank.

*We have four more variables, namely **hydraulic power, water inlet flow, wind turbine, electrical power** and **wind speed** (see* Figure 14.10).

Finally, the volume of water is emptied by the outlet flow which corresponds to the water consumption of the inhabitants.

*We can add the **output water flow** as an additional variable.*

Figure 14.11 shows the diagram of causal loops with the set of 11 variables that have been identified with their relationships. The causal analysis was realized around the volume of the reservoir.

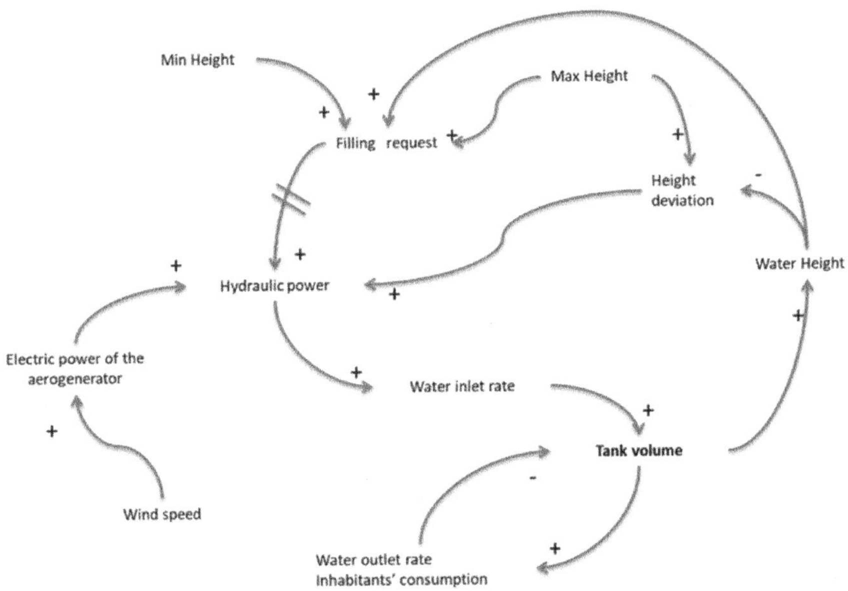

Figure 14.11. *Causal diagram for the mission "fill the water tank with the power of the wind"*

Among the remarkable elements, we have:

– A delaying relationship between filling demand and hydraulic power. Indeed, the regulation of the filling of the tank is done up to Hmax, but is triggered when the height of water is below Hmin. So, nothing happens between Hmax and Hmin.

– A loop between the volume of the tank and the water outlet flow. Indeed, the flow rate negatively influences the volume of the tank. The more the consumption increases, the more the volume of water in the tank decreases. We also have a positive relationship between the volume of the tank and the outlet flow. Indeed, the larger the volume and the greater the height of water and therefore the higher the output flow will be.

– A link between the wind speed and the electrical power of the wind turbine.

– The control of the system is done by the variable "filling request", which depends on the "height of water" and the limits "height min" and "height max".

– The regulation of the system, which is done by the variable "height difference", which is the difference between the "max height" setpoint and the observation of the water height.

14.5. Formalizing the simulation model

14.5.1. *Classification of variables*

The formalization of the qualitative model resulting from causal analysis consists of reclassifying the different variables. Forrester (1968) proposed a simple and effective method to facilitate the identification of the system of differential equations for simulation.

In system dynamics, there are three types of variables (see Figure 14.12). First of all, we have the **stock variables** or levels. These are variables that characterize the state of the system and its environment. For simulation purposes, the level represented by a cloud means an infinite source or an infinite well. Then, the **flow variables** represent the variations in levels. A flow variable involves time; it is a fraction of the level as a function of time. Time can be hours, minutes, seconds, as well as years, months or days. Finally, **auxiliary variables** can be combinations of variables, nonlinear expressions or simple constants.

Variables can be connected to each other with information flows represented by dotted arrows. For Forrester, the behavior of a system can be entirely characterized by its stock and flow variables. The identification of stocks and flows to characterize system behavior in its environment is part of the principles of "system thinking" (AFIS 2014).

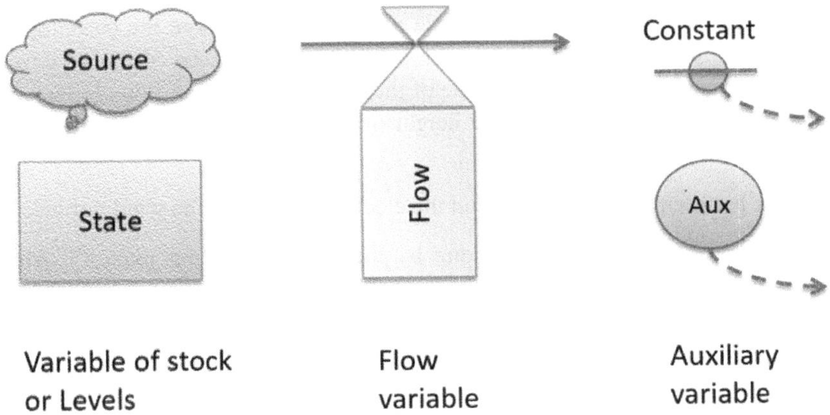

Figure 14.12. *Types of variables in system dynamics*

14.5.2. *Dynamic modeling*

For modeling system dynamics, Forrester relies on Norbert Wiener's cybernetics-defined feedback loop, adding an intuitive parallel with hydraulics. To format the classified stock, flow and auxiliary variables, Forrester proposed graphical modeling that gave rise to several computer languages including the DYNAMO language for "Dynamic Modelization" (Richardson and Pugh 1981; Sicard 1984). This language facilitates the formalization of equations and allows the execution of the simulation.

In Figure 14.13, we have a feedback loop with a flow variable which is the input action that will vary the state. This action is controlled by a decision that is adjusted by the state information.

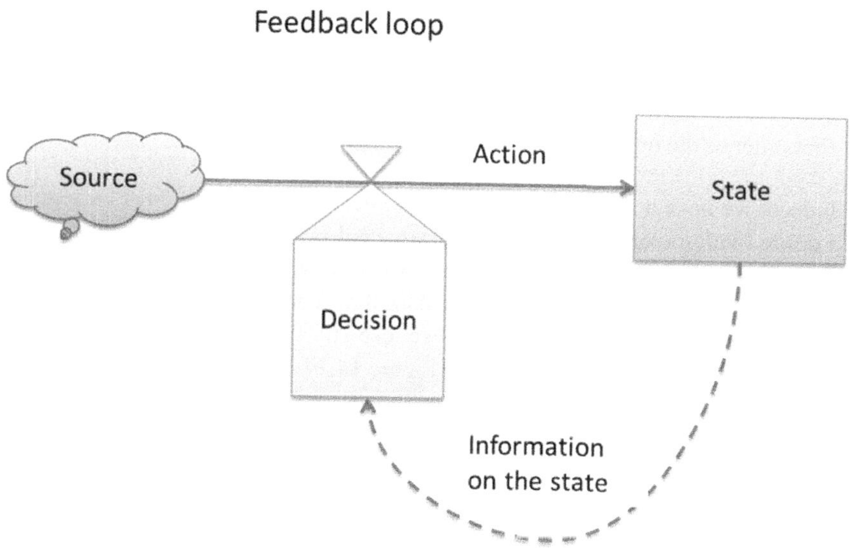

Figure 14.13. *First-order feedback loop*

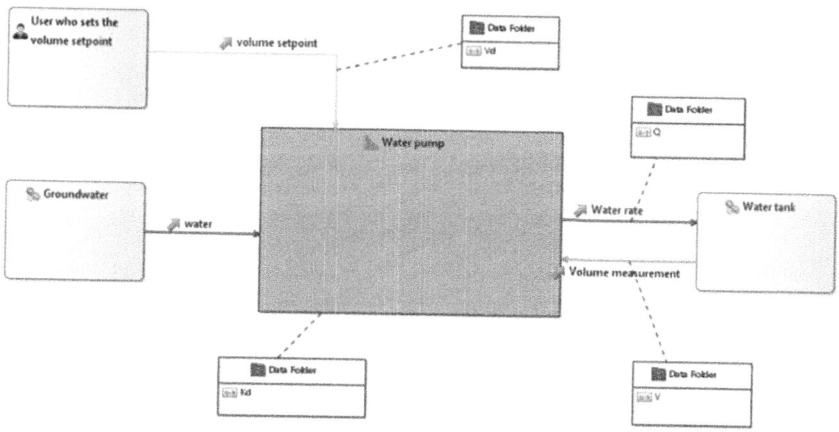

Figure 14.14. *Operational architecture of a water pump*

Let us take the example of a regulated pump system whose operational architecture is modeled in Figure 14.14. In the center, we have the black box pump system; on the right, we have the water table, which is a water resource for the system; on the left, we have the water tank, which is the external system "consumer" of the value produced by the pump, namely a water flow. Below, it is us, the stakeholder, who sets the filling instruction, namely a desired volume of water.

In Figure 14.15, we have the representation in the form of a negative feedback of the first order of the operational architecture of the water pump.

Indeed, we have a stock V, which is the water reserve, which must be regulated on a desired volume of water Vd. The variable Kd is a fraction of the stock per unit of time. The action Q, which is the input water flow, is calculated by the relation Kd*(Vd-V). In the example, Kd is a constant that represents a characteristic of our pump. To define this feedback loop, we use all the variables that have been identified on the operational modeling in Figure 14.14.

We can see the result of the simulation in Figure 14.15, with an initialization of the stock at 235 m^3 and a deposit at 300 m^3.

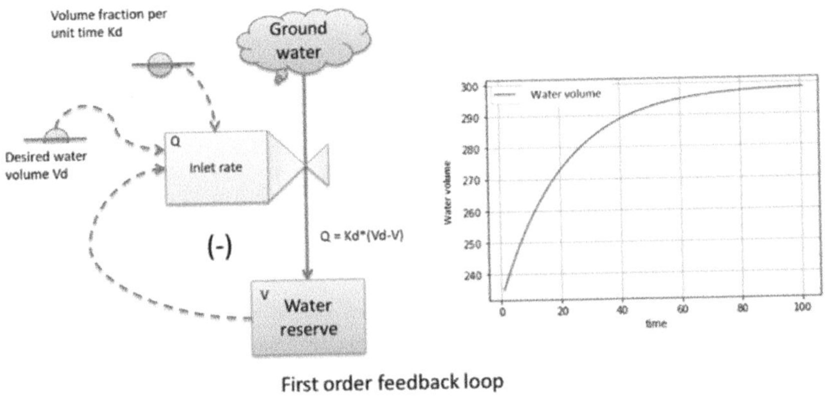

Figure 14.15. *Example of a negative feedback loop*

The goal of dynamic modeling is to convert the causal model into a formal stock and flow model. For this, we take as input the causal analysis from which we classify the different variables and we organize them.

The stock and flow diagram is built around the level variables (= stock variable). We then identify the flow variables and finally connect the auxiliary variables with information flows in order to calculate the flow variables. We finalize the diagram with missing information flows, for example, on stocks.

14.5.3. Application to the unifying thread example: formalization of the causal model

We resume the causal analysis of the mission to fill the tank. For the construction of this first simulation, we will make the following assumptions:

– For the stock variables, we will only take the volume of the reservoir with as input flow variable of the water flow of the water table, which is the output of the studied system (green frame in Figure 14.16), and as output flow variable of the flow of water from the consumption of the inhabitants influenced by the height of the water in the reservoir.

– We will consider as constant the wind as well as system parameters as for the wind turbine and the pump. Indeed, we consider at first that the transformation of the wind into useful power is instantaneous.

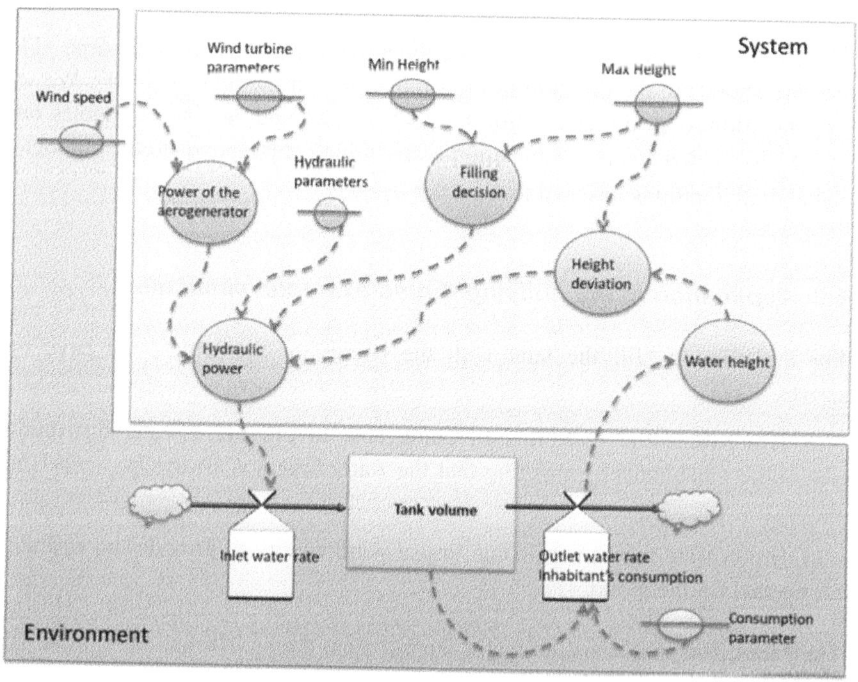

Figure 14.16. *Stock and flow diagram*

Figure 14.16 represents the formalization of the causal diagram of the mission "fill the reservoir" in a stock and flow diagram with the assumptions we have considered.

14.6. Running the simulation

14.6.1. *Identifying and configuring equations*

To run a simulation, we need to transform the stock and flow diagram into a system of differential equations translated into a computer language. There are many languages and software tools for simulation. In this book, in order to remain generic, we will directly use a translation of the equations in Python to test our different scenarios. Transforming the stock and flow diagram into equations requires further characterization of the different variables. However, the classification of variables in stock, flow and auxiliary form allows us to have an initial structure of the system of equations.

14.6.2. *Configuring the simulation*

To run a simulation, we need to define the initial time "T0", the final time "TF", which corresponds to the duration of the simulation and the time step "dt". We will call a simulation loop, the calculation of stocks, flows and auxiliary variables on a time step "dt". The number of simulation loops, which corresponds to the number of computational iterations, is equal to $Nbr = \frac{(TF-T0)}{dt}$.

14.6.3. *Application to the unifying thread example: simulation*

For the mission "fill the tank with the energy of the wind", we have two objectives:

– to demonstrate the feasibility of being able to provide a water distribution service to the inhabitants by ensuring that the water height is always between Hmin and Hmax;

– to demonstrate that it is possible to use wind energy to provide the hydraulic power needed for the pump.

The parameters that characterize the environment are as follows:

– the wind speed V in m/s is first assumed to be a constant between 3 and 8 m/s. For the simulation in Figure 14.17, we took a wind speed of 4 m/s;

– the volume of tank S in m^3 has a maximum capacity of more than 300 m^3;

– the consumption of water by the inhabitants is approximately 7 m^3 per hour. The outlet flow of the tank is a function of the water height.

The parameters that characterize the system are as follows:

– control parameters in meters Hmin = 3 m and Hmax = 5 m;

– diameter of the wind turbine in meters D = 9 m;

– electrical power of the wind turbine in Watt: Pelec = 0.29*D^2*V^3 Watt;

– pump parameter: we have a saturation at 55 m^3/h.

The initial parameters of the simulation are as follows:

– the initial time T0 = 0 and the final time Tf = 100 in hour;

– the time step is dt = 0.01 hour.

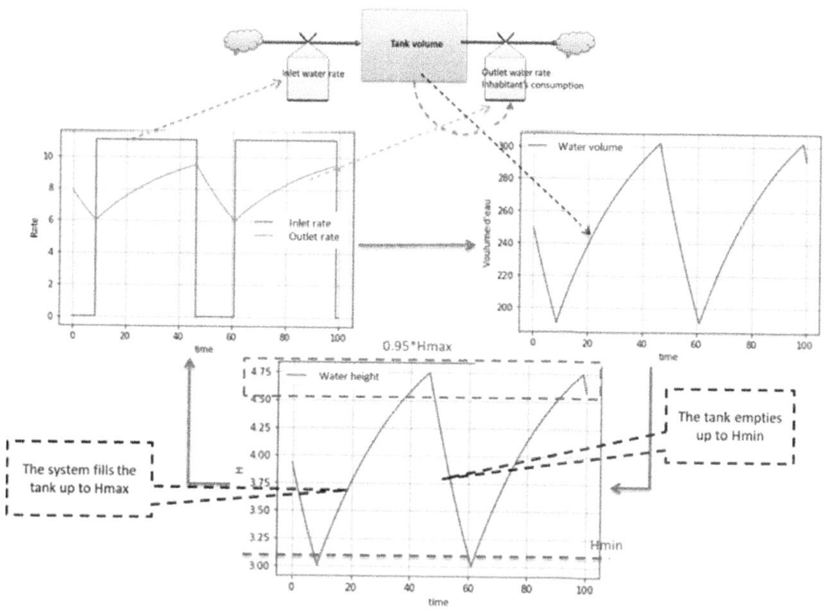

Figure 14.17. *Simulation result*

Figure 14.17 shows the simulation results for a tank with an initial water volume of 250 m^3 and a wind speed of 4 m/s. The water consumption of the inhabitants fluctuates between 6 and 9 m^3/h depending on the height of water in the tank. Indeed, the flow is greater when the water height is at its maximum. For the control of the system, the tank empties up to Hmin = 3 m and fills up to 95% of Hmax = 5 m using the power provided by the wind turbine.

14.7. Analyzing the results

14.7.1. *Purpose of the analysis of the results*

The analysis of the results is done from the outputs of the simulation in the form of graphs. The simulations being temporal, we will have in abscissa the time and in ordinate the observed variable or variables. By analyzing the responses of the system in its environment, we can look for remarkable points such as saturations or response times. For the verification of properties or the validation of operational scenarios, we can call on expertise. The business engineer or system architect analyzes the results as a result of the simulation. We will see that to facilitate this analysis, we can couple to the behavioral simulation model a property monitoring model.

14.7.2. *Application to a unifying thread example: analysis of the results*

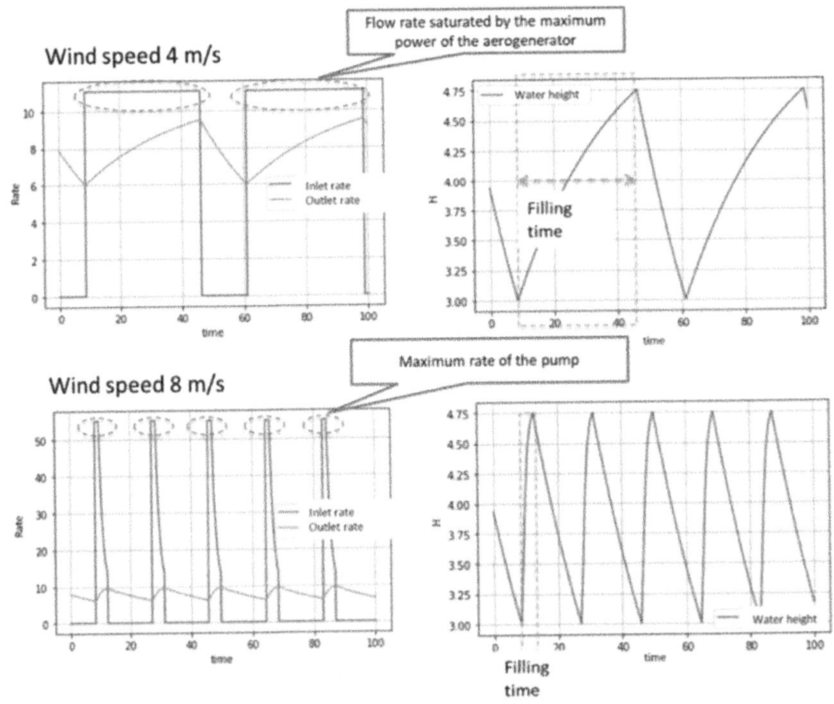

Figure 14.18. *Several filling scenarios depending on the wind*

Figure 14.18 shows us two results of filling the tank by changing as an initial condition the wind speed. As we could have intuitively deduced, the higher the wind speed, the faster the tank fills up. However, we have the saturation of the pump that comes into play for the scenario with 7 m/s of wind.

14.8. Conclusion on the realization of a simulable design model

A simulable design model is a simulation model that can have granularity ranging from the behavior of the black box system to the behavior of its equipment and the behavior of its subsystems.

The simulation of a design model must first and foremost meet a purpose, for example, development, verification of properties or the validation of operational scenarios.

In this chapter, we have presented a generic approach to the realization of design simulation models.

This approach builds on Donnadieu and Karsky's systems process as well as the dynamics of Forrester's systems.

The reader remains free to adapt this approach to their field or to apply directly, if they exist, the simulation techniques specific to their profession.

The results of a simulation are always in graphical form in order to allow for analysis and exploitation of the results.

This analysis step calls on the expertise of the trades or the system engineer and requires the formatting of the results to be properly adapted to what we are trying to observe.

15

Making a Simulable Specification Model

15.1. The monitoring model

15.1.1. *Why make a monitoring model?*

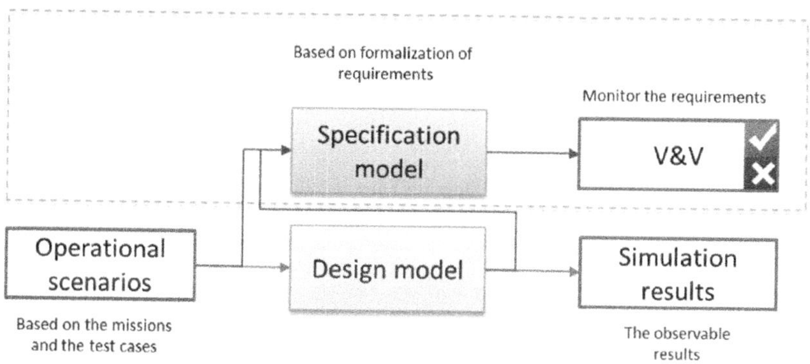

Figure 15.1. *Monitoring properties*

The advantage of simulation is that it enables the engineer to develop or verify properties and validate the feasibility or the proper realization of operational scenarios. Simulation results are rich in information and require proper formatting to allow engineers to perform their analyses.

The completeness of this formatting is not guaranteed, so we can use a properties monitoring model, coupled with the behavioral simulation model.

For a color version of all the figures in this chapter, see www.iste.co.uk/hygounenc/system.zip.

In Figure 15.1, the behavioral model is a design model that takes input stimuli from the operational scenario. The monitoring model is called the specification model because it verifies the requirements of the system. It takes inputs from the operational scenario stimuli and outputs from the design model. This will make it possible to verify the veracity of the monitored properties independently of the formatting of the simulation outputs.

15.1.2. *How to build a monitoring model*

The input to the process of building a monitoring model is the specification of the system requirements with its architecture model (see Figure 15.2). From there, if we have not already done so, we will resume writing the requirements in a structured form.

Next, we will make the link between these structured requirements and the causal loop model by identifying the variables that are in the requirements. From there we will be able to turn our requirements into properties.

This list of properties allows us to build a monitoring model based on a set of Boolean tests.

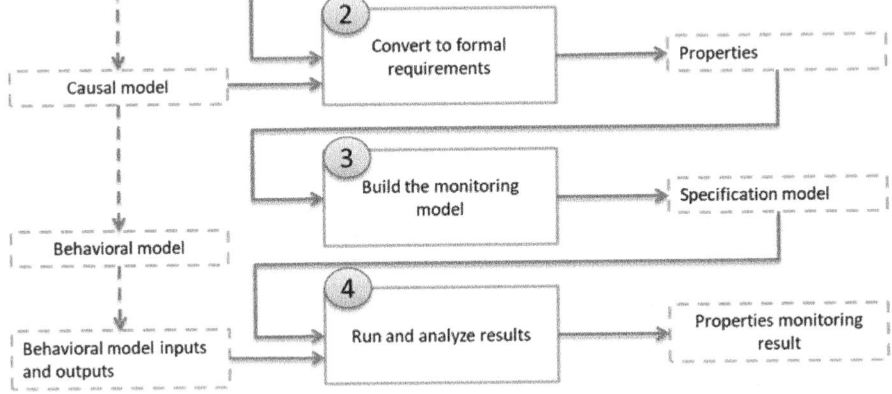

Figure 15.2. *Process for creating a properties monitoring model*

The monitoring model is our specification model containing the properties we want to verify on our simulable design model. The monitoring model is connected to the inputs and outputs of the design executable model.

The result takes the form of a list of properties that can take three values, namely "untested requirement", "tested and true requirement" or "tested and false requirement".

15.2. Writing structured requirements

15.2.1. *Reminder on the structure of a requirement*

In the section on the definition of requirements, we presented models for formatting functional and non-functional requirements. This semi-formal approach consists of writing the requirement in a structured form.

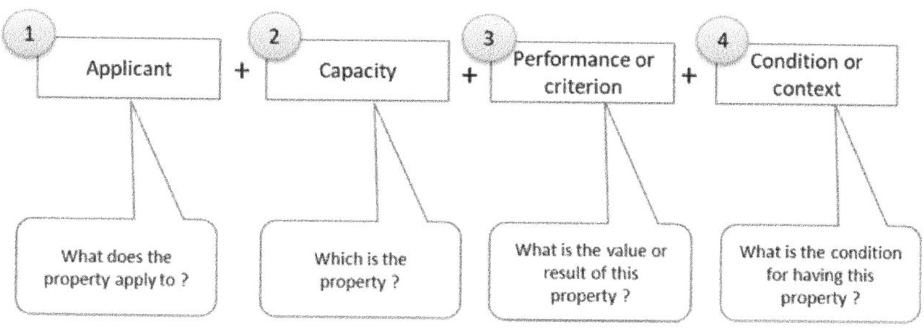

Figure 15.3. *Structured requirement*

A structured requirement can be a stakeholder requirement, or a system or subsystem requirement. It has four parts. First of all, there is the applicant who is either an element of the environment or the system, one of its functions or one of its components. Then we have the capacity which is the property described by the requirement. This property can be an ability to **do things** or to **be something**. Then there is the desired performance which is the expected value of this requirement. Finally, we have the condition that specifies the context in which this property is expected. We can find in Figure 15.3 the composition of a structured requirement.

15.2.2. *Application to the unifying thread example: selecting structured requirements*

For implementation on the unifying thread example, we will complicate the scenario by seeking to test the limits of the system. For this scenario, we will seek to fill the tank with wind whose speed changes every 24 hours from 8 m/s to 3 m/s.

The objective is to verify the following requirements:

Requirement 1: *the system shall fill the tank to 95% of the maximum height with a flow rate greater than 2 m^3/h, at the flow rate of water consumption in the "empty tank and wind" mode.*

The "empty tank and wind" functional mode is characterized by the following requirement:

Requirement 2: *the system switches to the "**empty tank and wind**" functional mode, if the wind speed is greater than or equal to 3 m/s, the tank has a water height less than Hmin and until the height is equal to or greater than 95% of Hmax.*

Requirement 3: *the system switches to the* **supply electricity to the grid** *mode, if the wind speed is greater than 3 m/s and the tank has a water height that can decrease by 95% from Hmax to Hmin.*

15.3. Formalizing property requirements

15.3.1. *Properties*

For Krob (2009), a requirement is a "logical" condition that must always be satisfied by the system to be made. This definition perfectly introduces the formal approach proposed by Micouin (2014) and implemented by Paper (2019).

For a classic V-cycle development, a requirement must be verified on the actual system. For a W development with virtual integrations, the idea is to put the requirement in the form of a partly Boolean property that can be verified on the simulated system and also on the real system.

Figure 15.4 presents a requirement in the form of a property that we will also call the **PBR**, "*property-based requirement*". This requirement model is divided into two elements. On the left, we have the logical condition that activates the property. This condition can be, for example, an operational context or a functional mode.

Then on the element on the right, we have the value of the property on a given object that must belong to a domain.

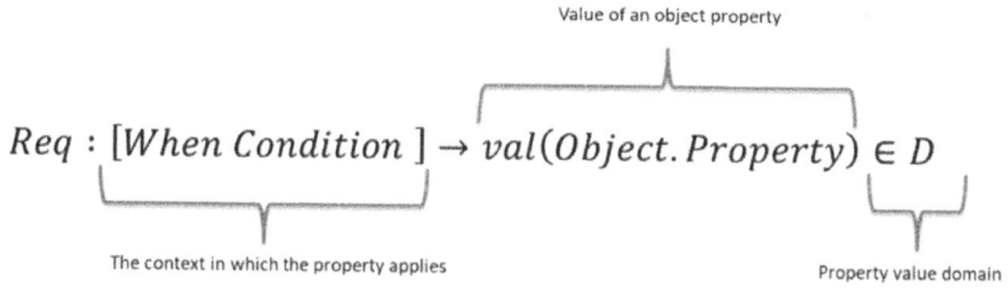

Figure 15.4. *Property or PBR (Micouin 2014)*

EXAMPLE.– the property can be the supply of a flow, an object, i.e. a pump. The supplied throughput must belong to a domain that is, for example, less than or equal to a maximum throughput.

15.3.2. *How to formalize a property*

To formalize our property requirements, we can use the structured form we have presented previously. As we can see in Figure 15.5, we can easily make the link between the structured requirement and the property.

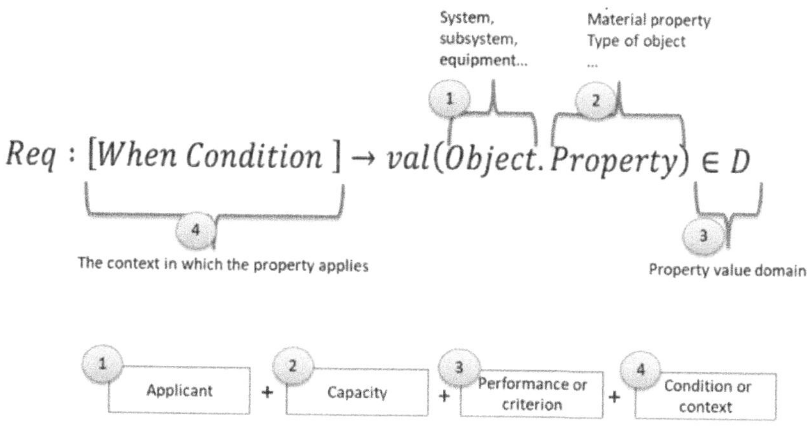

Figure15.5. *Relationship between property and requirement*

Indeed, the applicant is the object of the property. Capacity is the property itself. The criterion is the set composed of the value and the domain in which the property must be. Finally, the context of the requirement is the condition of ownership.

To be able to connect the properties of the specification model with the design model, we use for formalization, the variables from the **causal analysis** (see Figure 15.6).

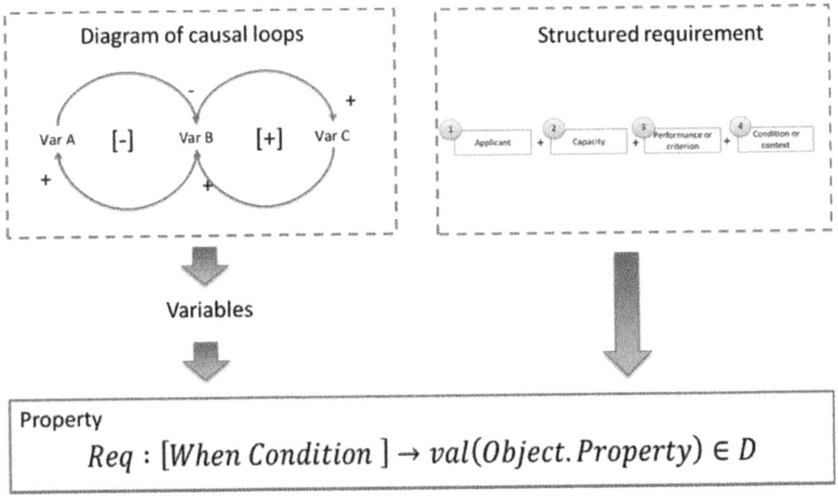

Figure 15.6. *Property with causal analysis variables*

15.3.3. *Application to the unifying thread example: formalization of properties*

We take the previous requirements and formalize them into properties using the variables identified as a result of causal analysis.

From **Requirement 1:**

"the system shall fill the tank to 95% of the maximum height with a flow rate 2 m^3/h higher than the flow rate of water consumption in the empty tank and wind mode"

We can write the **PBR1** property using the variables:

– "tank inlet flow", which is the "flow provided by the system";

– "reservoir outlet flow", which is the consumption of the inhabitants.

PBR1:

$$[\boldsymbol{When}\ (mode\ empty\ tank\ and\ wind)] \rightarrow System\ rate\ of\ flow \geq consumption\ rate\ of\ flow + 2\ m^3/h$$

We have the variable "full tank and wind mode" which appeared as an intermediate variable allowing a more concise condition to be obtained.

From **Requirement 2:**

> *"The system switches to the **'empty tank and wind'** functional mode, if the wind speed is greater than or equal to 3 m/s, the tank has a water height less than Hmin and until the height is equal to or greater than 95% of Hmax"*

We can write the **PBR2** property using the following variables:

– wind speed;

– H, the height of the water tank;

– Hmin, the minimum height of the water tank;

– Hmax, the maximum water height.

PBR2:

$$[\boldsymbol{When}\ (Wind\ speed\ \geq 3\ m/s)\ \boldsymbol{AND}\ (H < Hmin)\ OR\ (grad(H) \geq 0\ AND\ H \leq 0.95Hmax)] \rightarrow system\ mode\ = empty\ tank\ and\ wind$$

We find the system variable "empty tank and wind mode", introduced in PBR1, which corresponds to both conditions: wind speed greater than or equal to 3 m/s, water height less than Hmin and until the water height exceeds 95% of Hmax.

We have the variable grad(H), which is the gradient of the water height that allows us to know if the tank is filling.

From **Requirement 3:**

> *"The system switches to the **supply electricity to the grid** mode, if the wind speed is greater than 3 m/s and the tank has a water height that can decrease by 95% from Hmax to Hmin"*

We can write the **PBR3** property with the following variables:

– wind speed;

– H, the water height of the tank;

– Hmin, the minimum height of the tank;

– Hmax, the maximum height of the tank.

PBR3:

$$\begin{bmatrix} \textbf{When}\ ((Wind\ speed \geq 3\ m/s)\ \textbf{AND}\ (Hmin < H \leq 0.95Hmax)) \\ \textbf{AND}\ grad(H) < 0 \end{bmatrix}$$

$$\rightarrow system\ mode\ = supply\ electricity\ to\ the\ grid$$

We are introducing a new variable, "supply electricity to the grid" mode, which corresponds to both wind speed conditions greater than 3 m/s and a water height that can decrease from Hmax to Hmin.

We also have a grad(H) variable, which is the gradient of the water height of the tank and makes it possible to know if the tank is emptying.

15.4. Building a monitoring model

15.4.1. *Application to a property*

The objective of this activity is to translate the "PBR" properties into a Boolean relation executable on any simulation tool (see Figure 15.7).

For a PBR, the simulable module is broken down into two blocks that correspond to two test steps.

The first is to assess the condition of the PBR. If the condition is true, then we move on to the next step. On the other hand, if the condition is false, then the PBR is considered "not evaluated".

The second is to evaluate the PBR which will be true or false depending on the result (see Figure 15.8). An important note is that these two evaluations are done at each calculation stage. This means that the PBR is evaluated over the entire simulation horizon. This temporal approach is very interesting because it makes it possible to check when the condition is good and whether the property is true throughout the validity of thc condition.

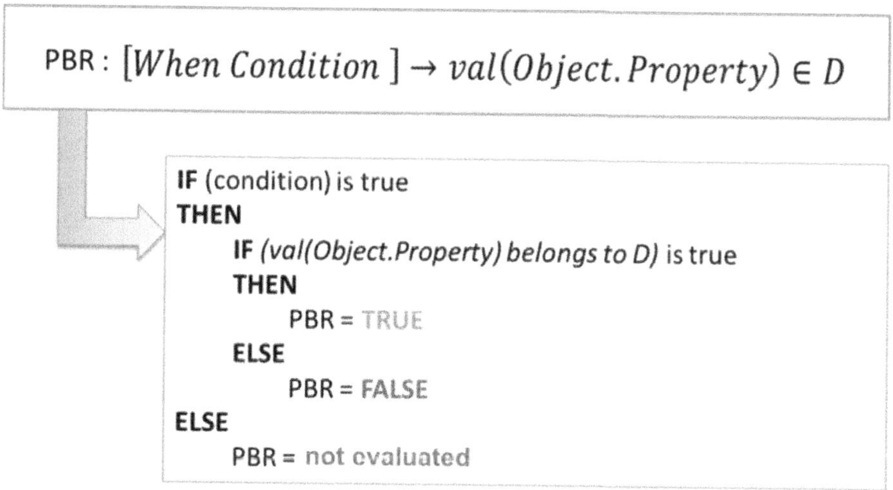

Figure 15.7. *Monitoring model for a PBR*

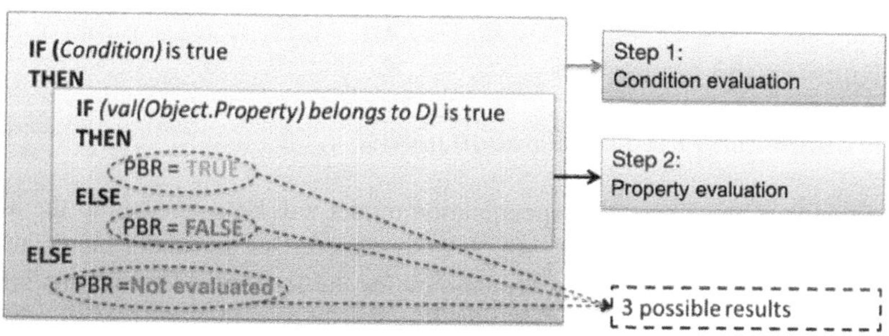

Figure 15.8. *Evaluating a PBR*

15.4.2. *Creating the specification model*

Each PBR must be transformed into an executable Boolean module. At the end of this activity, we gather all executable PDRs to form an executable specification module (see Figure 15.9).

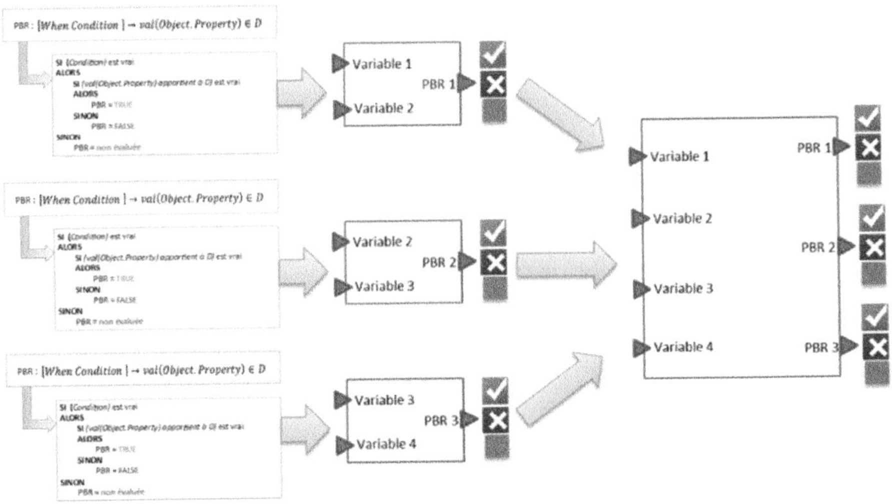

Figure 15.9. *Specification or monitoring model*

PBRs can use common variables. As an input to the executable specification model, we have a pooling of these common variables. On the other hand, at the exit, we find all the PBRs that we want to evaluate on our simulation.

15.5. Running the simulation

15.5.1. *Connecting the specification model*

The inputs of the executable specification model will be connected to the inputs and outputs of the design, model, which is simulated. Design model inputs are the variables used for the operational scenario or for the test case. The outputs are the results of the observed variables.

For each time step, the PBR properties contained in the specification model will be evaluated. At the output of the specification model, we will have all the evaluations of all PBRs over the time horizon of the simulation (see Figure 15.10). For each PBR and at each moment of the simulation, we will have three possible results, namely: the property has not been evaluated because the condition is not valid, the property is true, i.e. correct, or the property is false.

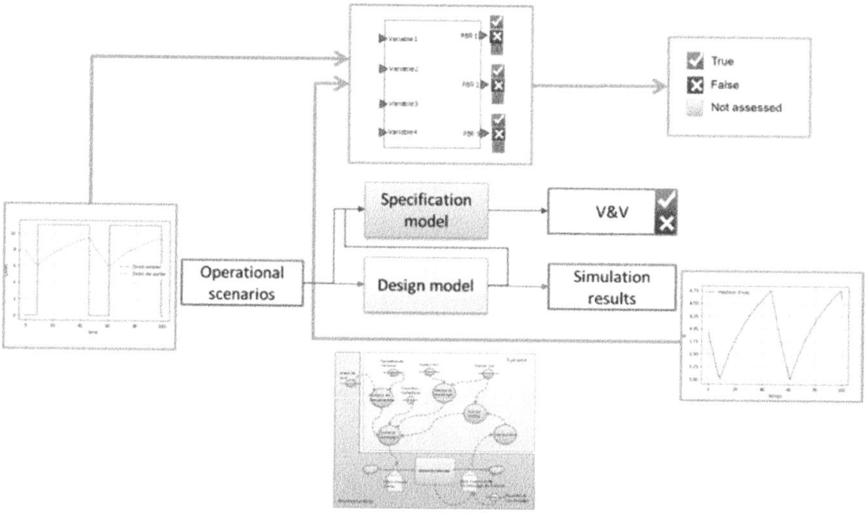

Figure 15.10. *Connecting the specification simulation model*

15.5.2. *Application to the unifying thread example: running the specification model*

To check if our three properties are present on our system, we will execute a scenario that allows us to look at how the system behaves with wind variations and water consumption that fluctuates depending on the height of water around a constant. The simulation will have a time horizon of 100 hours with an increment of 0.01 hours, which represents 10,000 calculation loops.

The volume of water from the reservoir is 250 m^3. The minimum height, Hmin, is 3 m, and the maximum height, Hmax, is 5 m. To stabilize the system, we will take 95% of this value, i.e. 4.75 m.

As input, we will vary the wind speed respectively 6 m/s, 4 m/s, 8 m/s and finally 3 m/s with an approximate period of 24 hours. The outlet flow of the tank that corresponds to the consumption will fluctuate between 6 and 10 m^3 per hour.

At the exit, we will observe the change in functional mode the height of water as well as the inlet flow of the tank in cubic meters per hour.

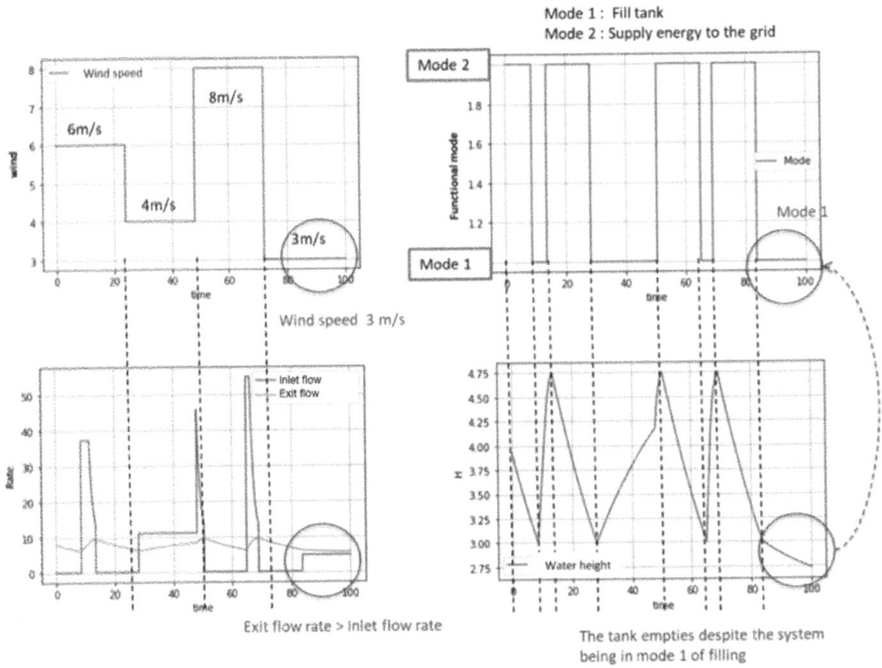

Figure 15.11. *Unifying thread example simulation filling of the tank with variable wind*

Figure 15.11 presents the results of the simulation with:

– wind variation top left;

– inlet and outlet flows of the tank bottom left;

– mode changes between filling the tank with wind energy and supplying electricity by routing wind energy top right;

– variations in the height of the tank bottom right.

We have annotated on the simulation results a remarkable point where, for a wind speed of 3 m/s, we observe an inflow lower than the outgoing flow, which has the consequence of emptying the tank while the system is in the filling mode. We will analyze these results through the simulable specification model with our three properties.

15.6. Analyzing the results

15.6.1. *Result from the specification model*

As with the simulable design model, the results of the specification model can be presented as a figure with time on the abscissa and the value of the PBR properties on the ordinate. For example, we can choose the following convention:

– if the PBR has not been evaluated, it is set to 0;

– if the PBR is evaluated and is false, it takes the value 1;

– if the PBR is evaluated and is true, it is set to 2.

This agreement may be amended at the discretion of the engineer. It can also be simplified. For example in Micouin (2014), the author proposes to consider an unvalued property as false.

15.6.2. *Application to the unifying thread example*

We will evaluate the three PBRs of the scenario specification model with wind fluctuations. In Figure 15.12, we have the results of the evaluation of PBR1, which as a reminder is:

> *PBR1:*
>
> $$[\boldsymbol{When}\ (mode\ tank\ empty\ and\ wind)] \rightarrow Flow\ rate\ of\ system \geq Consumption\ flow\ rate + 2\ m^3/h$$

This property states that the system must always provide the tank with a flow rate that is at least 2 m^3 per hour higher than the output flow corresponding to the consumption of the inhabitants, if the system is in the operating mode of filling the tank.

This property responds in particular to the need to fill the tank in less than 36 hours. The exogenous variable that allows us to see the effect of this property on the environment is the height of the water in the tank. First of all, we will check the conditions for activating the PBR1 evaluation.

We can see in Figure 15.12 that PBR1 is not well evaluated when the system is in "supply electricity to the grid" mode. As a reminder, during this functional mode, the tank empties from the height Hmax to Hmin, which allows the system in the presence of wind to sell the electricity produced.

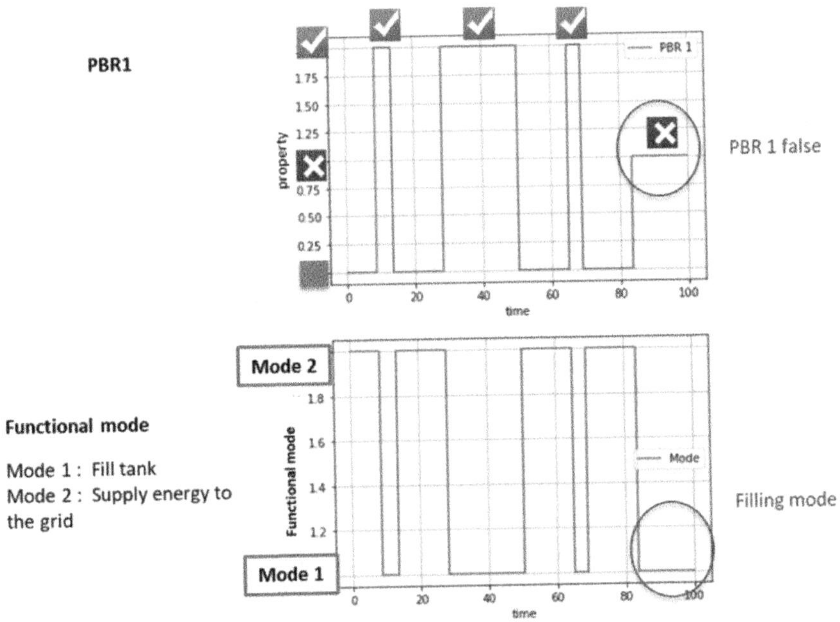

Figure 15.12. *Unifying thread example PBR1 evaluation*

We can also see that PBR1 is well evaluated when the functional mode of the system is "fill the tank with the wind". Of the four evaluations, we have three successes and one failure at the end.

In Figure 15.13, we can analyze the reasons for the successes and the failure. Whenever PBR1 has been successfully evaluated, this means that the incoming flow is greater than 2 m^3 per hour more than the outflow of consumption. In this case, the system fills the tank as we can see with the increase in water height. In the last case where PBR1 was evaluated as a failure, this means that the flow deviation is not the correct one. Indeed, we can see that the incoming flow is insufficient because the height of water decreases. The tank is emptying!

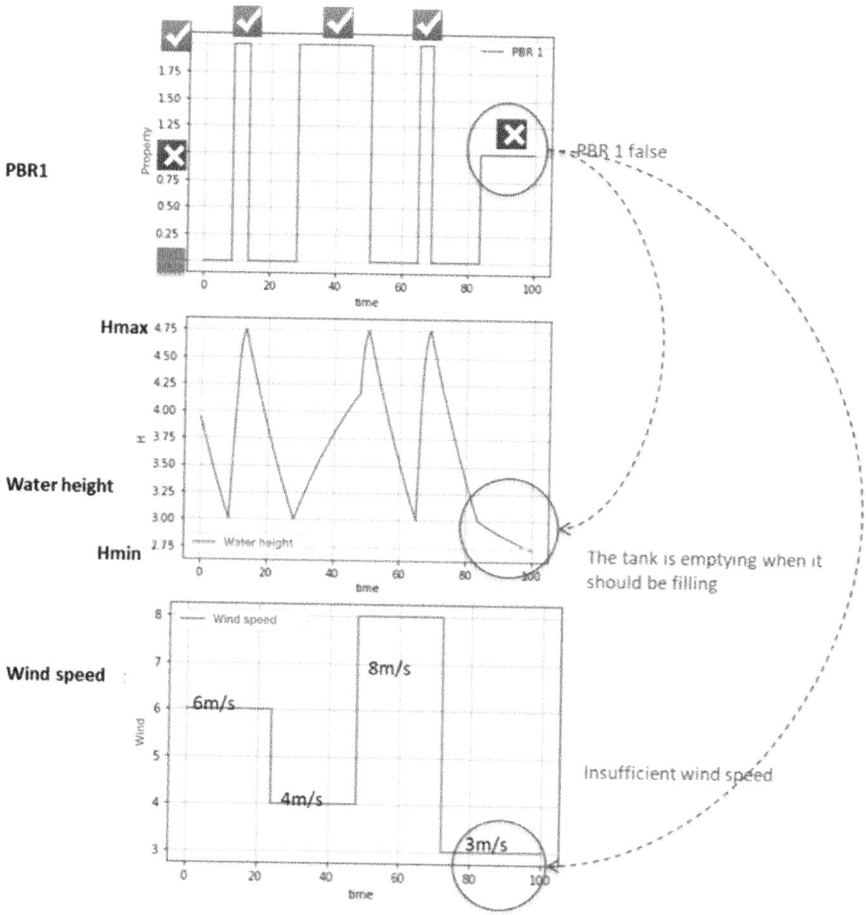

Figure 15.13. *PBR1 evaluation analysis*

The flow rate is induced by the hydraulic power of the pump which is powered by the electrical power of the wind turbine. We can deduce from this that it is insufficient. Indeed, the wind speed is only 3 m/s. Since the system must work for wind speeds between 3 m/s and 8 m/s to supply the power to fill the tank, we have two possibilities:

– renegotiate a compromise between needs and requirements in order to reduce the operating range of the system with wind and allow the power grid to take over;

modify the system design by taking up a wind turbine with a larger diameter.

We see that these two choices are not only purely technical. They must be studied in their entirety in order to identify and analyze the impacts in terms of performance, risks and costs.

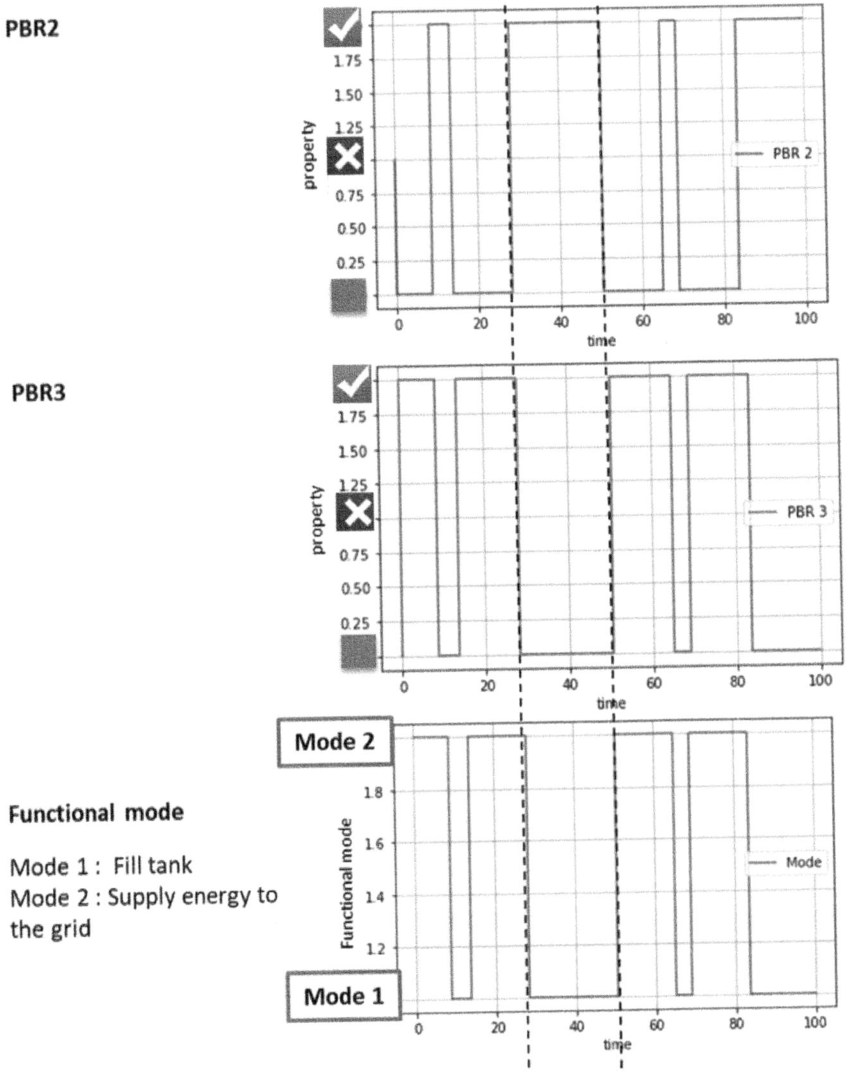

Figure 15.14. *Unifying thread example evaluation PBR2 and PBR3*

Figure 15.14 presents the results of the evaluations of PBR2 and PBR3. These two properties relate to the rules for moving the system from one functional mode to another.

As a reminder, the PBR2 property is formalized as follows:

PBR2:

$$[\boldsymbol{When}\ (Wind\ speed\ \geq 3\ m/s)\ \boldsymbol{AND}\ (H < Hmin)\ OR\ grad(H) \geq 0]$$

$$\rightarrow system\ mode\ = empty\ tank\ and\ wind$$

This means that for a sufficient wind speed and with insufficient water height or with a positive or zero height gradient (the tank fills up or remains at a constant height), the system switches or remains in the "fill tank with wind" mode.

As a reminder, PBR3 is formalized as follows:

PBR3:

$$\begin{bmatrix} \boldsymbol{When}\ ((Wind\ speed \geq 3\ m/s)\ \boldsymbol{AND}\ (Hmin < H \leq 0.95Hmax)) \\ \boldsymbol{AND}\ grad(H) < 0 \end{bmatrix}$$

$$\rightarrow system\ mode\ = supply\ electricity\ to\ the\ grid$$

This means that for a sufficient wind speed and with a water height between Hmax and Hmin and with a negative height gradient (the tank empties), the system switches or remains in the functional mode "supply electricity to the grid".

From the results of Figure 15.14, we can see that when PBR2 is evaluated as true, then the system is in functional mode "fill the tank with wind".

We can see the same with PBR3. When it is rated at "true", then the system is in "supply electricity to the grid" functional mode.

In addition, we can see that when PBR2 is evaluated, PBR3 is not and vice versa. This behavior is quite normal because when the conditions of PBR2 are met, those of PBR3 are not and vice versa.

15.7. Conclusion on the realization of a simulable specification model

The realization of a simulable specification model is done in several stages. First, we need to select the requirements we want to check.

Then, the systems engineer must resume the drafting of these requirements in a structured form if this was not already done. From there, it is possible to formalize the requirement in the form of a verifiable property.

This property can be easily transformed into a monitoring model in the form of logical tests.

For this monitoring model to be usable in a simulation, it is imperative that it shares the same variables as the design model and the operational scenario.

There are several advantages to this approach. The first is that the verification of properties by the monitoring model no longer depends on the formatting of the results of the design model. The second is that the execution of the monitoring model is repeatable on several simulations or even on the test bench.

PART 6

System Integration

16

Using Models

16.1. Models in systems engineering

In this chapter, we discuss the use of models in systems engineering. The model of a system is the description of that system from a certain point of view (Krob 2017). That is, the model has only certain properties of the system and only the properties that are present on the model can be studied. A model is a representation that can be of different natures. We can have (AFIS 2012):

– **concrete models**, for example, a model made in 3D printing to study physical properties;

– **abstract models** like the kinematic equations of a vehicle;

– **hybrid models** involving concrete and abstract models as well as real elements of the system. For example, a HIL (hardware-in-the-loop) simulation, where we can have the ECU with the embedded software as the real part of the system, the power supply as a concrete model and the environment and the rest of the system in the form of equations as an abstract model.

16.2. Typology of models

In Le Moigne (1999), we can only understand and grasp a complex system through modeling. A purely analytical or reductionist approach would have the effect of simplifying the system and causing the loss of interactions at the origin of emerging behaviors. We will see that there is a wide variety of models and that some are more "disabling" than others in restoring the properties of the system.

For a color version of all the figures in this chapter, see www.iste.co.uk/hygounenc/system.zip.

Systems Engineering,
by Emmanuel HYGOUNENC.

Figure 16.1 presents a representative panel of models that we can use in systems engineering (AFIS 2012). We have the family of cognitive models that make it possible to model the system or its environment in order to understand, to explore or to formalize knowledge.

We have the system models for analysis and design. These are descriptive models based on semi-formal languages that make it possible to prescribe and architect a system and its environment.

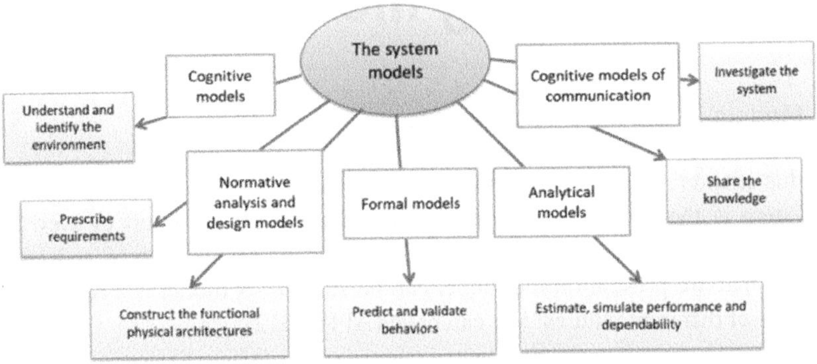

Figure 16.1. *Typology of models*

We also have formal models, based on mathematical relationships that make it possible, for example, to provide evidence of system behavior or to verify system properties.

Finally, we have the analytical models which we will present in a little more detail. We have two types of analytical models. We have behavioral models and business models (Krob 2017).

16.3. Analytical behavioral models

Analytical behavioral models rely on the representation of states (see Figure 16.2). For a system, the vector of states represents the set of internal variables that characterize its dynamic behavior over time.

To make the link with the systems and the modeling approach by the dynamics of systems of Jay Forrester (Forrester 1968), the state variables are the variables of level or stock. As with the dynamics of systems, we start from the assumption that the system is **causal**, i.e. its future evolution depends only on the past, and

deterministic, i.e. the initial conditions determine the future of the system. The representation of states allows for modeling of the functional behavior of the integrated system and its components.

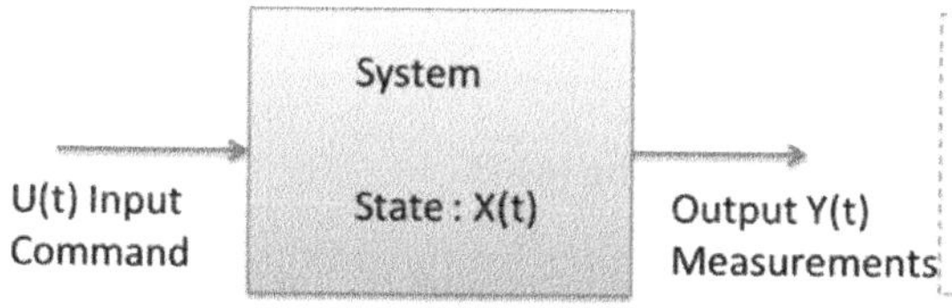

$$\dot{X}(t) = F[t, X(t), U(t)]$$
$$Y(t) = H[t, X(t), U(t)]$$

Figure 16.2. *Status representation*

16.3.1. *Using the behavioral model in V&V*

The behavioral analytical models will allow for verification of the functioning of the system components and the functioning of the integrated system. Indeed, these models are well suited to model the behaviors of the components and integrate them together to obtain the overall behavior at the system level.

In addition, behavioral analytical models can be used to model the behavior of the environment as needed. By using the operational scenarios as input to the behavioral model of the integrated system, it is possible to carry out its functional validation.

In the virtual integration presented in Chapter 13, we saw that it was possible to carry out several levels of verification and validation depending on the granularity of the integrated models.

16.3.2. *Model-based design*

Model-based design uses analytical models in the development, verification and validation of systems.

This design approach is incremental. If all the built-in elements of the system are behavioral models, we will talk about **model in the loop**. It is this approach that we use for the virtual integration process (see Figure 16.3).

We will talk about software-in-the-loop (SIL) if the embedded software is "one for one" and if the rest of the system components, as well as the environment, are simulated (see Figure 16.4).

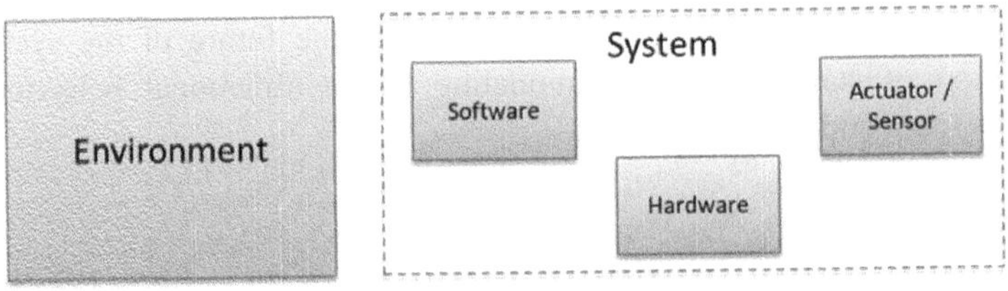

Model In the Loop (MIL)

	Model	Real
Software	X	
Hardware	X	
Actuator / Sensor	X	
Environment	X	

Figure 16.3. *Model in the loop*

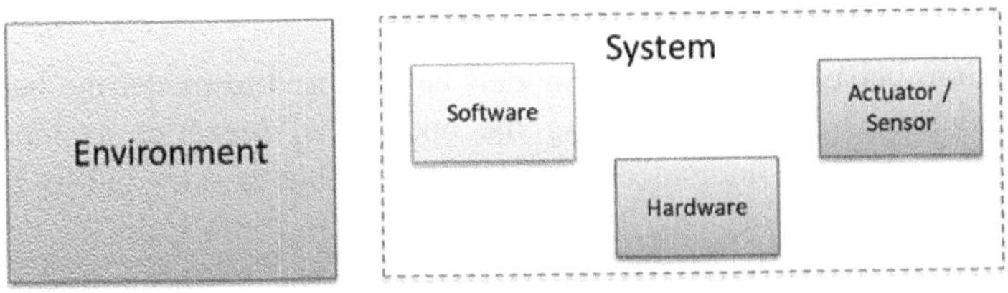

Software In the Loop (SIL)

	Model	Real
Software		X
Hardware	X	
Actuator / Sensor	X	
Environment	X	

Figure 16.4. *Software-in-the-loop (SIL)*

The next integration phase consists of verifying and validating the embedded software loaded into its ECU. In this case, we are talking about hardware in the loop (HIL) (see Figure 16.5).

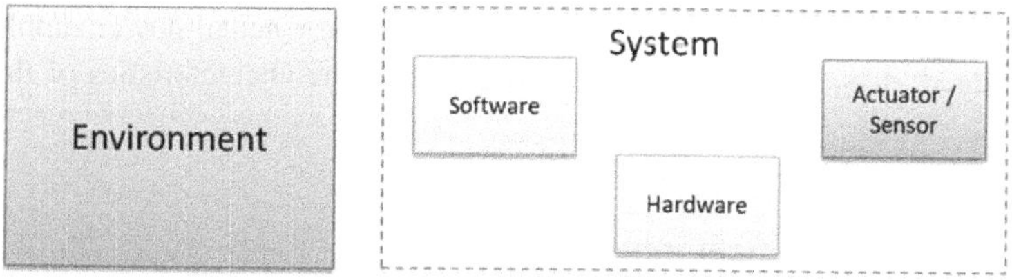

Hardware In the Loop (HIL)

	Model	Real
Software		X
Hardware		X
Actuator / Sensor	X	
Environment	X	

Figure 16.5. *Hardware-in-the-loop (HIL)*

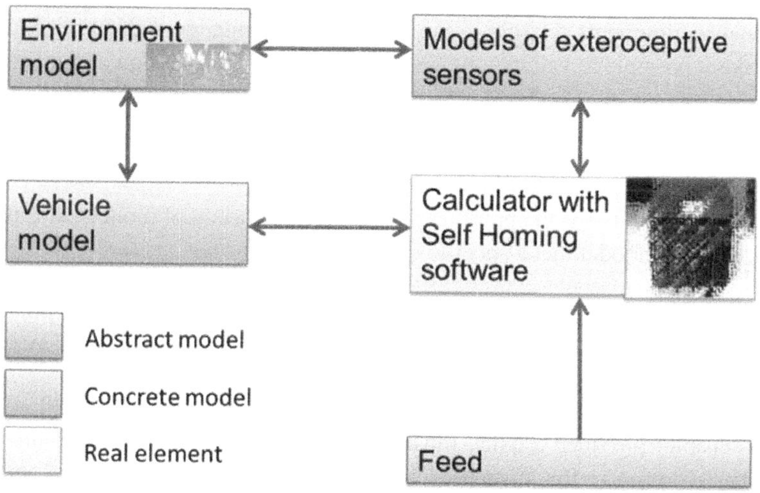

Figure 16.6. *Example of HIL for a drone*

In Figure 16.6, we have an example ofhardware-in-the-loop, with the ECU and autoguidance software that are real, and the sensors and actuators of the vehicle that are modeled in an abstract way just like the environment. We will note that the power supply of the calculator is a concrete model of the actual power supply of the drone. Indeed, it is a laboratory power supply with the characteristics of the drone battery.

16.4. Business analytical models

"Business" analytical models are views of modeling the system from the angle of a business specialty also called "engineering", for example, "mechanical engineering". This business view for studying the properties of the system is necessary, but not sufficient, because it is too reductive or "disabling". These models can be used for system sizing, verification and validation. Business models seek to be as close as possible to the laws of physics related to the profession. They make it possible to study scenarios representative of reality, but in a limited state of the system.

16.4.1. *Using the business model in V&V*

Business analytical models make it possible to verify the functional and non-functional requirements relating to the business area under consideration. Similarly, they make it possible to validate the needs relating to the profession in question.

16.5. Conclusion on the use of models

We discovered in this chapter that there is a wide variety of models. We also saw that each type of model met a specific objective.

We were also able to discover the MBD (model-based design) approach and understand the enrichment brought to this V&V approach with the use of monitoring models presented in Chapter 15.

This chapter provided a quick illustration of the main incremental techniques of model-based design, for example, hardware-in-the-loop, where the embedded electronics and software are real and the rest of the system and environment are simulated.

We also presented the business models with their advantages and limitations.

17

System Integration, Verification and Validation

17.1. System integration

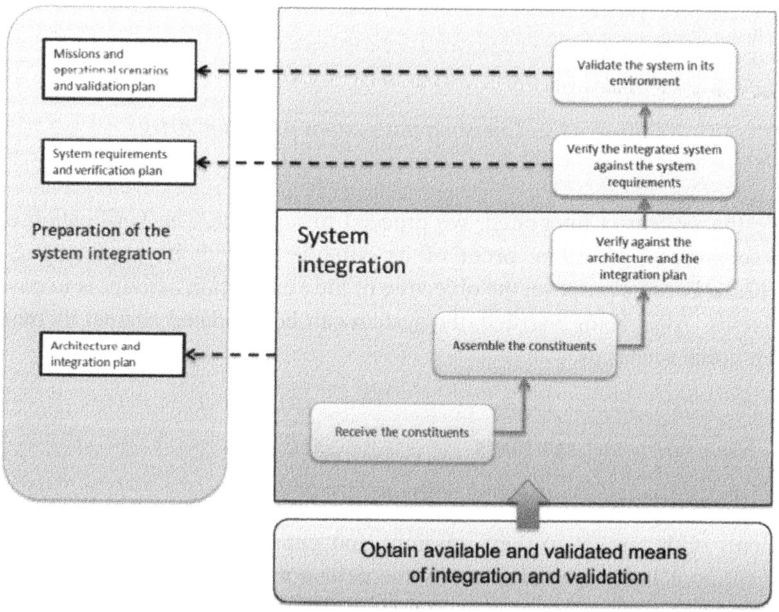

Figure 17.1. *System integration (Faisandier 2017)*

For a color version of all the figures in this chapter, see www.iste.co.uk/hygounenc/system.zip.

The aim of the integration activity is to provide our integrated system with its various components as output. Integration consists of receiving the various constituents and validating them in relation to their respective missions. Then, we assemble the different components by testing each of the intermediate integration steps (see Figure 17.1).

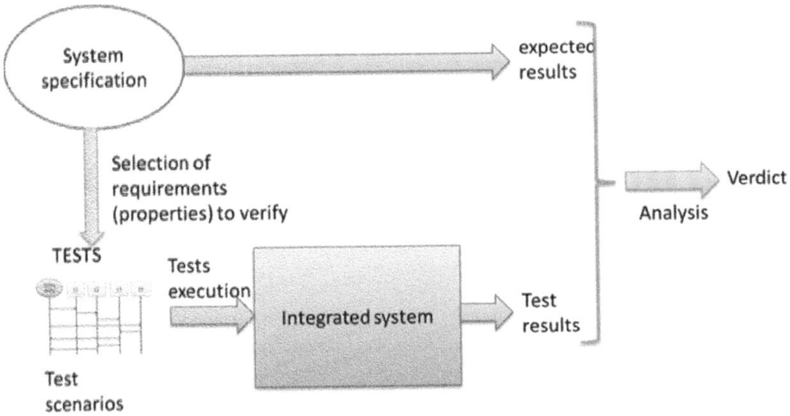

Figure 17.2. *Checking the system (Longuet 2016)*

Once the system is integrated, we proceed to verify it. The verification of the system consists of providing proof of its compliance with its requirements (see Figure 17.2). For this purpose, the objective of the verification activity is to cover all requirements. After being verified, the system can be validated against its missions and operational scenarios.

17.1.1. *Receiving subsystems*

The purpose of receiving subsystems is to validate the compliance of the constituents with respect to their missions and compliance with their constraints. This reception activity is carried out by the person responsible for the system to be integrated. Each component or subsystem supplier provides the integrator with the V&V and compliance matrices with respect to the requirements allocated to the subsystem. The compliance matrix is the supplier's response to the system owner's initial requirements. The V&V matrix, also called the justification matrix, is the set of proofs of verification of the component or subsystem against its specification. These various elements are validated by the person in charge of the system to be integrated.

17.1.2. *Assembling subsystems*

The assembly of the subsystems is done by following the system integration plan. The integration plan describes the strategy chosen to integrate the various subsystems. It also describes engineering activities related to integration, organization, resources and means (Faisandier 2017). The integrator writes the integration procedures based on the definition of interfaces and the design of the system to verify:

– the functional between the different subsystems (the interfaces);

– the physics of each subsystem (e.g. mass, dimensions, color, etc.);

– the start of the global functioning (link with the verification of the final system).

The integration procedures are carried out. Results and corrective actions are documented in accordance with the V&V plan.

17.2. Checking the system

Product verification consists of checking if our system has all the properties that are found in the specification. We are trying to answer the question: Did we make a good system? That is, does our system work as we have specified?

Figure 17.3 provides a process for verifying the integrated system based on the EIA-632:1999 standard (ANSI 1999). To verify the system, two activities need to be completed:

– **Preparing the verification**: this activity requires the verification to be broken down into the different procedures according to the methods and means of testing. This is an important preparatory step that gives rise, in particular, to the drafting of a verification plan.

– **Conducting the verification**: this verification activity is carried out by following the procedures from the verification plan.

Depending on the results and after analysis, part of the verification may have to be carried out again by adapting the procedures in order to make it possible to conclude on the conformity of the system with respect to the requirements. The final results will be recorded in a verification analysis report and anomaly or non-compliance sheets.

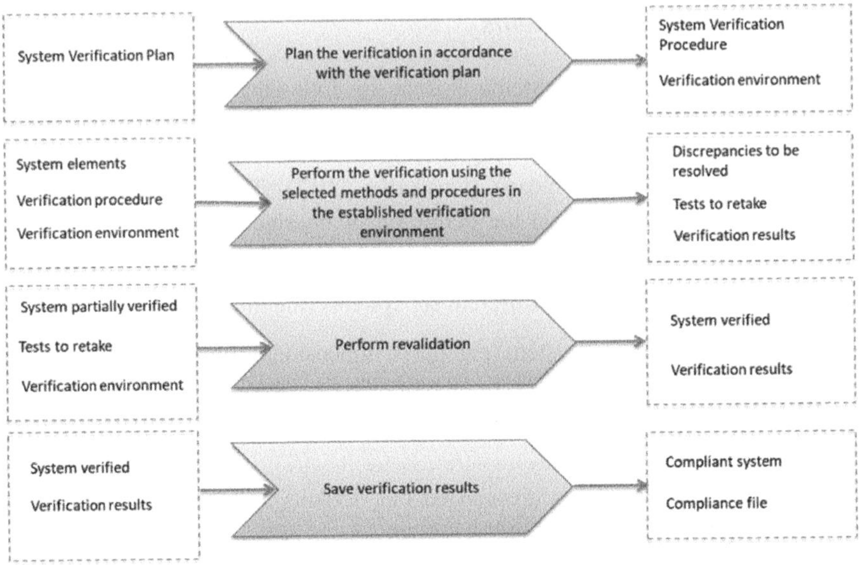

Figure 17.3. *Checking the system and its components*

17.2.1. *Defining the system verification strategy*

The system verification strategy is initiated at the definition of system requirements and must be enriched throughout the development cycle. The verification strategy consists of listing the principles adopted in terms of:

– verification method;

– distribution of verification actions during phases and milestones of the project;

– systems contributing to the verification;

– distribution of verification actions between systemic levels.

Once the methods are identified, we refine the allocation on test means for tests and simulation means for simulations. The objective is to control the cost of the verification exercise and to be able to plan the availability of test means, which requires knowledge of the duration of the tests.

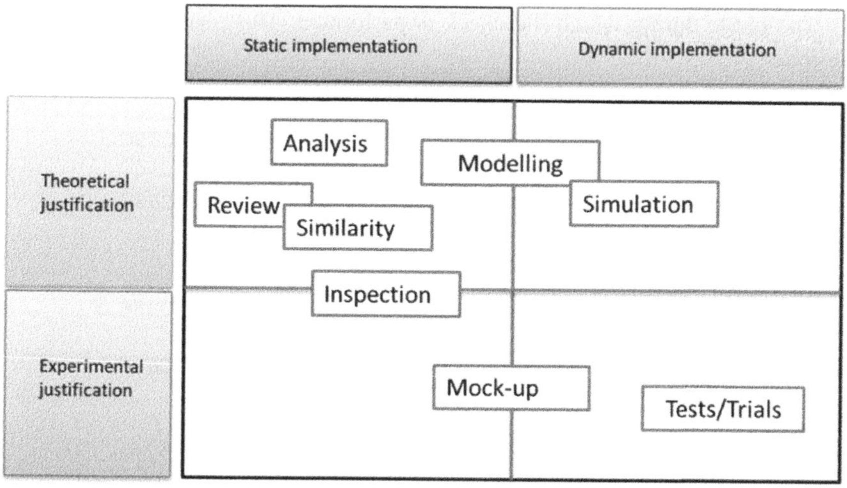

Figure 17.4. *Verification and validation methods (Faisandler 2017)*

Figure 17.4 presents the main methods used for the verification. They are divided into two groups:

– methods that are implemented statically, i.e. without the obligation to have the functioning of the system;

– dynamic verification methods that require the system to work.

These two verification approaches provide justifications that can be theoretical or experimental. As an annex to the V&V plan, a justification matrix (also called a V&V matrix) must be established at the outset of the definition of the system requirements framework. For each requirement, we must provide at least the following information:

– The verification (or validation) action to ascertain the presence of the property. We need to build at least one test for each requirement and conversely know the requirements covered by each test (see Figure 17.5).

– The object on which the action is applied.

– The verification method used, for example: analysis, modeling, inspection, similarity, testing. Verification methods can be combined, for example analysis and testing.

– The phase during which the check is performed. This phase can be a project milestone.

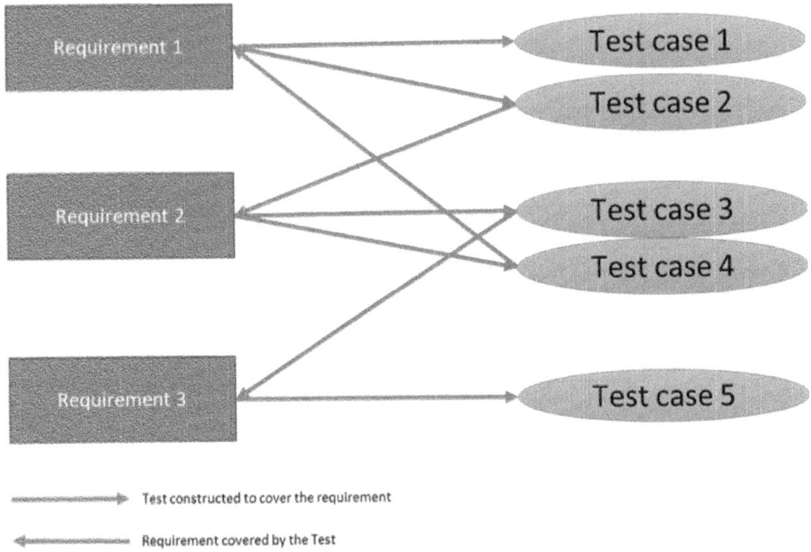

Figure 17.5. *Splitting and grouping tests (Longuet 2016)*

17.2.2. *Defining the system verification procedures*

The verification procedures define how the system will be checked against its requirements. A verification procedure must include the following preparatory information:

– the verification action concerned or the verification actions concerned;

– the necessary contributing systems: test means, simulator, etc.;

– the configuration of the object concerned by the verification action.

We need to describe the execution of the verification procedure which consists of a set of tasks. In the procedure, it is particularly necessary to give the initial conditions, describe the tasks to be done and also describe the conditions for success, i.e. the expected results.

17.2.3. *Performing the system check*

The verification activity concerns the **functional** and **physical** aspects of the system, as well as the **dysfunctional** aspects. The verification activity should cover all requirements of the specification. The objective of the verification is to cover all requirements (see Figure 17.6).

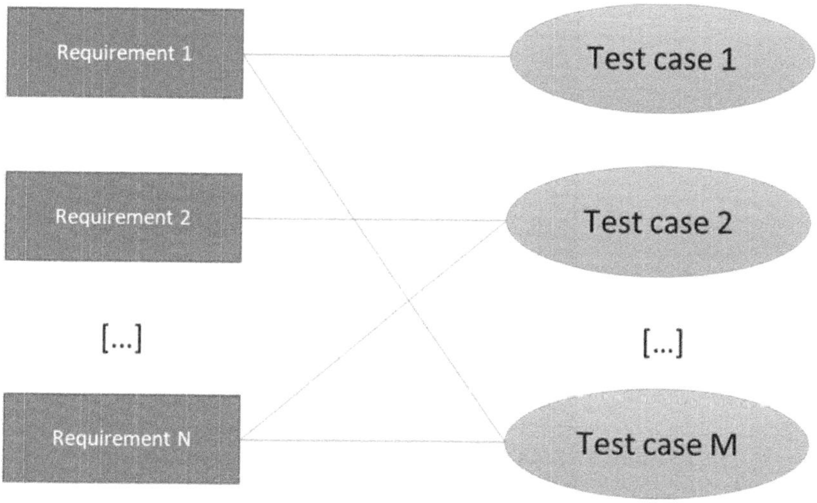

Figure 17.6. *Requirements coverage (Longuet 2016)*

When drafting requirements, it is important to think about verification methods. At the same time, it is possible to initiate the drafting of the verification plan. The verification action is implemented by the selected method and is carried out by the verification procedure. The verification action takes as input a referential that can be one or more requirements of the system (or subsystem) and an object that can be a function, a component or the system. As an output, we describe the expected result (see Figure 17.7).

At the end of the execution of the verification procedures, we will record the results obtained. Depending on the expectation, the results can be "pass" or "fail". The results of the checks are analyzed to see if they are valid or invalid. Valid procedures whose results are "pass" are declared compliant. Valid procedures with "fail" results must be analyzed to determine whether they are compliant or non-compliant. The invalid procedures must be repeated and a reverification will have to be carried out.

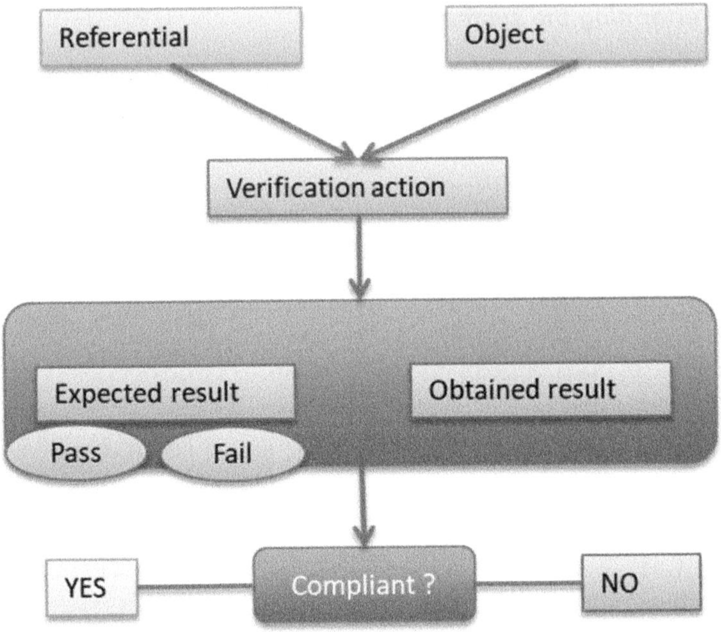

Figure 17.7. *Verification action (Faisandier 2017)*

17.2.4. *Establishing system compliance*

The purpose of recording the results is to establish proof of the conformity of the system with respect to its properties (requirements). A verification report is required to see compliance with requirements. Operational safety studies need to be updated with the actual behavior of the product. If the product is compliant, it can be delivered to be validated. If there is non-compliance, then the design or specification may be questioned. Proof of system compliance should enable us to **make the link** between product, requirements and needs.

17.3. System validation

Product validation consists of carrying out operational tests of the integrated system in its environment, in order to validate the response to the needs. We are looking to answer the question: Did we make the right system? This activity is carried out by the project owner (contracting authority) in its test environment or in an operational manner. The activity can be carried out to a certain extent by the project manager (prime contractor) with partial representation of the operating environment. In this case, it is often combined with the verification activity. Or the activity can be carried out jointly between the contracting authority and the prime contractor. In this case, we can talk about a recipe test.

In Figure 17.8, we have a validation process inspired by the standards EIA-632:1999 (ANSI 1999) and ISO/IEC-15288:2015 (ISO 2015). The process inputs define the types of validation we want to do (operational validation, contractual validation, certification). At the output, we have the validated system with the proofs that must be recorded such as the report and the anomaly sheets.

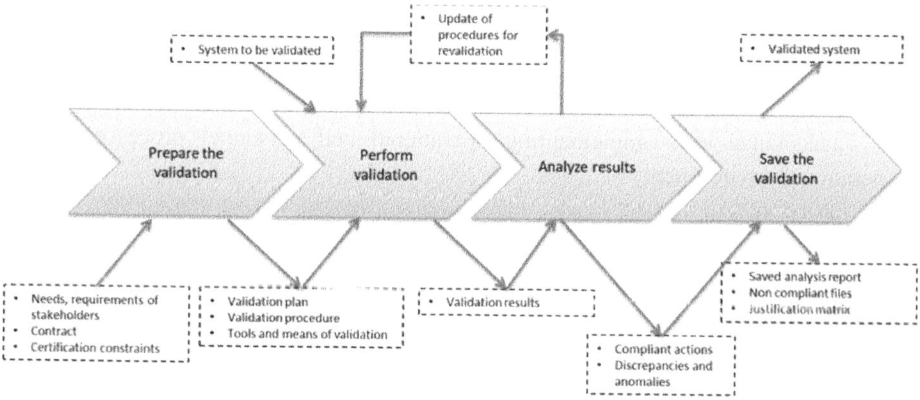

Figure 17.8. *Validation process*

The implementation of a validation process can take place in the following ways, drawing inspiration from the standards EIA-632:1999 (ANSI 1999) and ISO/IEC-15288:2015 (ISO 2015):

– define the validation strategy;

– set up the validation environment and perform the validation;

– identify compliance and deviations as well as invalid procedures;

– perform revalidation for invalid validation procedures;

– record results and prepare data for corrective actions.

17.3.1. *Preparing for validation*

The organization of system validation is defined in a validation plan. The final system validation plan defines:

– the validation strategy of the system with the types of validation, the actions and the methods to be applied;

– the validation environment with the necessary means, tools and resources;

– the elements of organization and planning of activities.

On the final system that needs to be validated, we need to determine the type of validation with the applicable output criteria and stakeholder requirements. Among the types of validation to be carried out on the final system, we have:

– validation against the requirements of user stakeholders in a test environment representative of operational conditions;

– certification tests against certification requirements;

– acceptance tests implementing operational and personnel processes in an operational environment;

– the tests or validation requests specified in the contract.

In a justification matrix, which is attached to the validation plan, we must associate with the stakeholder, certification or contractual requirements, a validation action and a validation method among those presented for verification.

17.3.2. *Performing validation*

To perform system validation, test procedures must be written according to the validation plan. The procedure explains step by step how the validation action takes place according to the selected method.

Figure 17.9 shows the progress of a validation action. As input we have the referential which can be, for example, stakeholder requirements and an object that can be the system, a subset of the system or a component of the system.

Then, we have the validation action with the expected result. From the result obtained during the validation procedure, we can calculate the compliance level, which is the deviation from the expected result. The validation criterion allows us to say whether it is acceptable or not. We need to justify the relevance of the results of this validation action.

Before carrying out the validation procedures, it is necessary to ensure that we have all the necessary tools and means. All the results produced must be analyzed before being recorded.

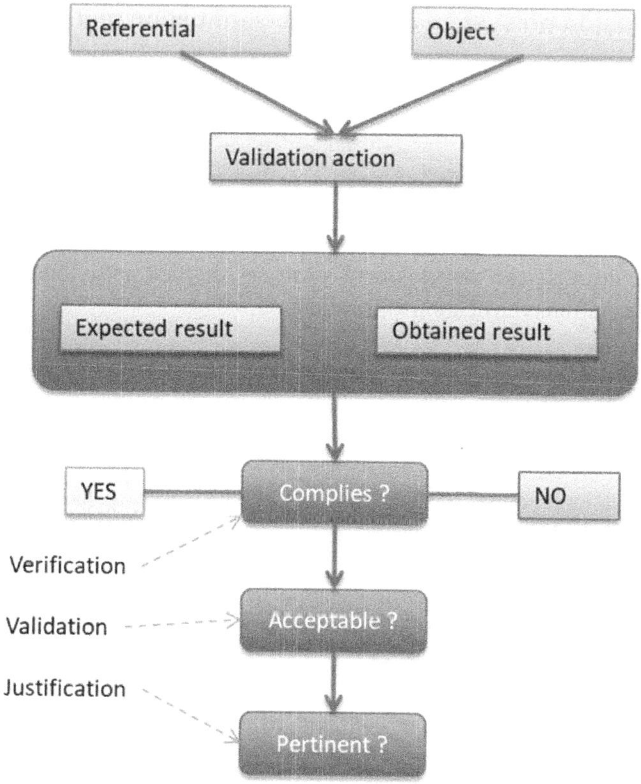

Figure 17.9. *Validation action (Faisandier 2017)*

17.3.3. *Analyzing the results*

The results are compared with the expected results. Gaps are analyzed with validation criteria. When a result is non-compliant and the test procedure is valid, it is an anomaly. When the result is consistent and the procedure is valid, the action is validated. Invalid procedures will have to be corrected, and a revalidation will have to be carried out.

17.3.4. *Saving validation results*

Validation results must be the subject of an analysis report. Each non-compliance is the subject of an anomaly sheet. The analysis report and anomaly sheets must be recorded by the project. The rationale matrix needs to be updated.

17.4. Conclusion on system integration, verification and validation

In this chapter, we have addressed the rise of the development cycle whose activities are common, regardless of the shape of the V- or W-cycle.

The W-cycle adds an activity specific to the recalibration of the models that makes it possible in particular to optimize the level of V&V testing to be carried out.

Rollback through the development cycle is divided into three system-level activities: integration, verification and validation.

Integration consists of validating the various components of the system with respect to their respective missions. The various components will have to be assembled and tested in order to obtain the integrated system.

The verification consists of checking if all the properties of the specification are indeed present on the integrated system.

And finally, validation makes it possible to confirm that the system responds well to its missions and respects all the constraints imposed by its environment. This activity is under the responsibility of the project owner, but can be shared with the project manager.

General Conclusion

This book is intended for engineers and students who wish to discover or apply systems engineering.

Throughout the sections and the different chapters, we have illustrated the main themes that cover the engineering of complex systems.

The **first part** allowed us to present the richness of systems theory. We have defined a system and in particular a complex system. The proverb that best describes a complex system is "**the whole is more than the sum of its parts**". This highlights the particularity of these systems, which is the emergence of properties. The conclusion of this part is that complex systems need to be approached comprehensively and for that they need to be modeled.

The **second part** introduced systems and requirements engineering. We presented the main processes that constitute the systems engineering approach. These processes can be classified into three families, namely the definition of requirements, the design of the system and cross-cutting processes such as V&V, the evaluation and comparison of candidate architectures. In this part, we also introduced the CESAM architectural framework that paces the various activities implemented in technical processes. We presented a technical and managerial process for requirements engineering. We have shown that these technical activities can be found in the various processes of systems engineering. On the other hand, the requirements management process was a crucial process that made it possible to establish a link with project management in particular.

In **part three** on defining requirements, we presented the activities of the needs analysis and requirements specification processes. We ended this part by showing how to validate a requirements framework.

Systems Engineering,
by Emmanuel HYGOUNENC. © ISTE Ltd 2023.

In **part four** on system design, we detailed the activities of the functional and logical design processes with its process of verifying and validating this abstract architecture. We then detailed the physical design activities with its process of verifying and validating this concrete architecture. We concluded this section by presenting the process for evaluating and comparing solutions.

In **part five** on virtual system integration, we developed the implementation of a development cycle in W by adding cycle escalations in the simulation. We presented a generic system simulation method applied to design models. We completed this simulation approach by presenting a method for observing system properties in the form of a specification model.

In **part six** on system integration, we presented, as an example, all the activities for integrating system components, verifying the properties of the integrated system and validating the system as part of a joint revenue activity between the system manufacturer and the client.

Finally, we recalled that a complex system must be approached in **its entirety**. To do this, it must be **modeled**. The system model consists of a black box model and a white box model.

The black box model represents the missions of the system, its network with its environment (it is a network of interacting systems) as well as its evolution through operational contexts.

The white box model is built in two steps: first from an abstract angle, where the engineer determines which system functions meet the missions. These functions are broken down into sub-functions until they can be allocated to physical components. These sub-functions form a functional network of interactions that evolves according to the functional modes. The second step consists of building a network of components that performs the functions and missions of the system. This network evolves according to technical configurations that implement functional modes and cover all the operational contexts of the system.

A complementary approach to this black box and white box modeling is the modeling of causal loops. The identification of exogenous and endogenous variables and their positive and negative interactions will give a model of causal loops. This model will allow the engineer to analyze the origins of complexity (feedback loops, delays, etc.) and build a behavioral simulation.

References

AFIS (n.d.) Vérification & validation tout au long du développement [Online]. Available at: http://ftardy.free.fr/FTP/Docs%20IVVQ/V&V%20dans%20le%20Cycle%20de%20developpement.pdf.

AFIS (2004). Glossaire de base de l'ingénierie de systèmes [Online]. Available at: https://homepages.laas.fr/kader/Glossaire_IS.pdf.

AFIS (2009). *L'ingénierie d'une ligne de produit*. Cépadues, Toulouse.

AFIS (2012a). *Découvrir et comprendre l'ingénierie système*. Cépaduès, Toulouse.

AFIS (2012b). *Guide bonnes pratiques en ingénierie des exigences*. Cépaduès, Toulouse.

AFIS (2014a). *Guide bonnes pratiques en expression du besoin*. Cépaduès, Toulouse.

AFIS (2014b). *Introduction au Penser Système*. CFC, Paris.

ANSI (1999). EIA standard: Process for engineering a system. EIA-632, January.

Aracil, J. (1984). *Introduction à la dynamique des systèmes*. Presses universitaires de Lyon, Lyon.

Audry, F. and Taillard, P. (2010). *La démarche d'analyse fonctionnelle*. Académie de Versailles, Versailles.

Baeckelandt, M. (1992). *Manuel d'ingénierie système. Comment utiliser les normes américaines pour le développement des systèmes complexes*. Masson, Paris.

von Bertalanffy, L. (1968). *Théorie générale des systèmes*. Dunod, Paris.

Boirel, R. (1988). Des interactions au système coexistant. *Revue internationale de systémique*, 2(3), 271–279.

de la Bretesche, B. (2000). *La méthode APTE*. Éditions Pétrelle, Paris.

Cressent, R. (2012). Valorisation de l'ingéniérie système à base de modèle, pour l'analyse de sureté de fonctionnement des systèmes complexes critiques intégrant des COTS [Online]. Available at: https://theses.hal.science/tel-00812898/document.

Donnadieu, G. and Karsky, M. (2002). *La systémique, penser et agir dans la complexité*. Editions Liaisons, Paris.

Duysinx, P. (2016). GEMMA. Paper, Université de Liège, Liège.

Eppinger, S.D. and Browning, T.R. (2012). *Design Structure Matrix Methods and Applications*. MIT Press, Cambridge, MA.

Faisandier, A. (2012). *Systems Opportunities and Requirements*. Sinergy'Com, Belberaud.

Faisandier, A. (2013). *System Architecture and Design*. Sinergy'Com, Belberaud.

Faisandier, A. (2017). *Evaluation et preuve du système*. Sinergy'Com, Belberaud.

Forrester, J.W. (1968). *Principles of Systems*. Pegasus Communications, Watham, MA.

Forrester, J.W. (2009). *Some Basic Concepts in System Dynamics*. MIT Press, Cambridge, MA.

Friedenthal, S., Moore, A., Steiner, R. (2009). *A Practical Guide to SysML*. Morgan Kaufmann OMG Press, Burlington, VT.

Galinier, M., Galinier, S., Laporte, C.Y. (2017). *Les Clés de l'ingénierie des systèmes*. Cépadues, Toulouse.

Garcia, J.M. (2019). *Modèles de simulation avec dynamique des systèmes*. Self-published, Spain.

Hoarau (2010). Adopter la pensée systémique et entreprendre efficace [Online]. Available at: www.pourquoi-entreprendre.fr.

Hull, E., Jackson, K., Dick, J. (2011). *Requirements Engineering*. Springer, Berlin, Heidelberg.

IEEE (2007). IEEE-1220. Systems Engineering – Application and management of the systems engineering process. Standard.

INCOSE (2015). *Systems Engineering Handbook: A Guide for System Life Cycle Processes and Activities*, 4th edition. Wiley, Hoboken, NJ.

ISO (2015). ISO/IEC-15288. System engineering – System life cycle processes. Standard.

Krob, D. (2009). *Eléments d'architecture des systèmes complexes*. CNRS Edition, Paris.

Krob, D. (2014). *Eléments de systémique – Architecture des systèmes*. CNRS & Ecole Polytechnique, Paris.

Krob, D. (2017). *CESAMES Systems Architecting Method*. CESAM Community, Paris.

Le Moigne, J.-L. (1983). La théorie du système général [Online]. Available at: https://entrorger-anndefrenne.com/wp-content/uploads/2021/04/LE-MOIGNE-J.-L._La-theorie-du-systeme-general.-Theorie-de-la-modelisation_.pdf.

Le Moigne, J.-L. (1999). *La modélisation des systèmes complexes*. Dunod, Paris.

Longuet, D. (2016). Vérification et Validation. Cours 4 – Introduction au test et méthodes informelles [Online]. Available at: https://www.lri.fr/~longuet/Enseignements/16-17/Et4-VV/Et4-4TestInformel.pdf..

Mathelot, P. (1988). L'Ingénierie des systèmes d'information. Groupe d'études et de recherches en management de l'information, Paris.

Meadows, D. (2008). *Thinking in Systems*. Earthscan, London.

Meinadier, J.-P. (1998). *Ingénieirie et intégration des systèmes*. Hermes Science-Lavoisier, Paris.

Micouin, P. (2006). Proposals for the definition and the implementation of systems engineering processes [Online]. Available at: https://pastel.archives-ouvertes.fr/pastel-00002518/document.

Micouin, P. (2014). *Model Based Systems Engineering*. ISTE Ltd, London and John Wiley & Sons, New York.

NASA (2007). *NASA System Engineering Handbook*. NASA, Washington, D.C.

Paper, P. (2019). How modeling and simulation methodology can introduce a real preventive approach in systems development. Event, MathWorks.

Pollet, Y. (2007). Intégration de systèmes à logiciel prépondérant [Online]. Available at: https://www.afscet.asso.fr/msc/textes-2008/Pollet-22oct07.pdf.

Pollet, Y. (2016). *Systèmes, architectures, intégration : pratique de l'intégration des systèmes et des logiciels avec SysML*. Éditions Ellipses, Paris.

ProjectSmart (2014). The Standish Group CHAOS. Report, ProjectSmart, Boston, MA.

Richardson, G. and Pugh, A. (1981). *Introduction to System Dynamics Modeling with DYNAMO*. MIT Press, Cambridge, MA.

SAE International (2010). Guideline for development of civil aircraft and systems. Standard ARP4754A. SAE International, Warrendale.

Sicard, G. (1984). DYNAMO, un compilateur pour parler comme le Club de Rome. IAE, DESS Microinformation et gestion, Lyon.

Stéphane Badreau, J.-L.B. (2014). *Ingénierie des exigences*. Dunod, Paris.

Tassinari, R. (2006). *Pratique de l'analyse fonctionnelle*. Dunod, Paris.

Thomé, B. (1993). *Systems Engineering Principles and Practice of Computer Based Systems Engineering*. John Wiley & Sons, New York.

Villemejane, J. (2013). Conception d'une machine à état synchrone [Online]. Available at: https://docplayer.fr/108259024-Conception-d-une-machine-a-etat-synchrone.html.

Voirin, J.-L. (2017). *Model-based System and Architecture Engineering with the Arcadia Method*. ISTE Press, London, and Elsevier, Oxford.

Weilkiens, T. (2014). *Variant Modeling with SysML*. MBSE4U, Fredesdorf.

Wiener, N. (1948). *La cybernétique, information et régulation dans le vivant et la machine*. Le Seuil, Paris.

Index

150% architecture, 196

A, B

abstract models, 291
actual integration of the final product, 245
allocated requirements, 56, 180
allocating system requirements, 52
analytical
  approach, 8–10
  models, 292
architectural visions, 47
assumptions, 146, 187
auxiliary variables, 261
basic requirement, 84
black box, 78
  point of view, 22
block diagram, 18
business models, 296

C, D

canonical forms, 84
causal
  diagram, 20
  link, 252
  modeling, 251
causal loops
  diagram, 259
CESAM architectural framing, 47
cognitive models, 292
combinatorics of environmental states, 91
complex system, 5, 10, 22
  model, 26
complicated system, 5
component
  of a system, 193
  tree, 191
concrete models, 291
configurations, 219
  management, 59
consistency of requirements repository, 147
constraints, 54, 89, 101
contributing systems, 106
control
  chains, 122, 125
  flow, 155
cost criteria, 237
coupling
  analysis of, 170
  matrix, 205
  quality of the, 206
cybernetics, 262
  system, 29
delay, 257
derived requirements, 146
design model, 247, 280

Design Structure Matrix (DSM), 168, 205
downward traceability, 145, 182
dynamic
  functional architecture, 154
  modeling, 252

E, F

emergence, 11
emergent properties, 12, 225
energy chain, 159
environment, 23, 74
event, 113
external
  functional interfaces, 123
  interfaces, 79
  physical interfaces, 124
feedback loops, 20, 22, 26, 83, 255
flow variables, 261
formal models, 292
functional
  analysis methods, 158
  behavioral architecture, 156
  chains, 117, 121
  context, 105
  design, 41, 153
  interface requirement, 138
  interfaces, 168
  requirements, 127, 128, 136
  scenario, 124
  vision, 30, 46
functioning of the system, 176
functions, 100, 116, 118, 119
  of the system, 158

G, H

global vision, 8
goal, 35
hardware-in-the-loop (HIL), 294
holistic vision, 8
hybrid models, 291

I, K

influence diagram, 20
information chain, 159
initial requirements repository, 56, 111
integration
  activity, 298
  plan, 299
interface matrices, 206
internal environment, 23
Kano diagram, 83
key factors, 257

L, M

life
  cycle, 87, 88, 105
    requirements, 137
  situations, 88
limiting factors, 257
logical components, 180
  architecture, 197
main goal, 91
managing change, 61
MEI (matter, energy and information)
  flow, 78, 117, 119, 120, 211
mirror model, 247
missions, 35, 38, 91, 114
modeling, 18
monitoring model, 248

N, O

needs, 39, 54
  analysis, 40, 71
  enchantment, 84
  requirement, 35
non-emergent properties, 12
non-functional requirements, 127–129
operating
  chains, 122
  context, 87
  modes, 113, 171

operation
chain, 126
of the functions, 176
operational
scenario, 96, 105, 114, 249
vision, 29, 45
optimization problem, 227
organic vision, 30, 46

P, Q

performances, 136
requirement, 84, 136
perimeter of the system, 79
PESTEL (political, economic, societal, technological, ecological and legal) analysis, 73
physical
architecture, 191
context, 105
decomposition, 202
design, 42, 191
functioning, 223
interfaces, 215
ports, 217
requirement, 139
scenario, 223
prime contractor, 41, 51
problem domain, 54
procedures for verifying the physical architecture, 232
product
breakdown structure, 209
tree, 201
validation, 304
verification, 299
project owner, 40, 51
property, 274
allocation, 228
based requirement, 274
monitoring model, 271
purpose, 35, 91
qualitative
model, 254, 261
modeling phase, 251

R, S

receiving subsystems, 298
relationship between variables, 255
requirements, 112, 126
configuration, 59
definition, 52
engineering, 51
activities, 55
management, 57
life cycle, 61
repository, 63, 111
specification, 41
validation, 56, 143
process, 144
version, 59
risk criteria, 237, 238
schematic diagram, 180
sequence diagram, 19
services, 35, 89, 100, 105, 119
simulable module, 278
simulation, 21, 251, 252
network of equations, 252
sizing of a system, 227
software-in-the-loop (SIL), 293
solution domain, 54
specifications, 102, 135
document, 134
model, 247, 272, 280
specified requirements, 56, 229
stakeholders, 72, 105
requirements, 36, 40
state variables, 292
static functional architecture, 153
stock
and flow diagram, 264
variables, 261
structured requirement, 273

sub-functions, 154, 157
system, 3, 35
  behavior, 124, 262
  behavioral model, 248
  compliance, 304
  dynamics, 261
  requirements, 126
    repository, 56, 111
  validation, 305
systemic, 3
  approach, 9, 251
  exploration, 15
    phase, 251
  models, 292
systems
  engineering, 36
  of systems, 6

T, U

technical
  configurations, 191
  functions, 157
TEF changes, 119
test case, 249
traceability, 61
types of properties, 224
upward traceability, 145, 182
uses, 38
  case, 93
  functions, 119
utilization case, 93

V, W

V-
  cycle, 50
  shaped development cycle, 48
validation, 43
  of functional architecture, 186
  of the physical architecture, 233
  process, 305
value criteria, 237
variables, 254
variants, 196
verification, 43
  activity, 303
  methods, 301
  of functional architecture, 185
  of the physical architecture, 231
  strategy, 300
virtual
  integration, 247
    of the system, 245
  lift, 245
W-cycle, 48, 50
weight of each criterion, 239
white box point of view, 22

Printed and bound by CPI Group (UK) Ltd, Croydon, CR0 4YY
13/09/2026
14960998-0006